Birds, Bones, and Beetles

Birds, Bones, and Beetles

The Improbable Career and Remarkable Legacy of University of Kansas Naturalist Charles D. Bunker

Chuck Warner

University Press of Kansas

Published by the University Press of Kansas (Lawrence, Kansas 66045), which was organized by the Kansas Board of Regents and is operated and funded by Emporia State University, Fort Hays State University, Kansas State University, Pittsburg State University, the University of Kansas, and Wichita State University.

This book is nonfiction. So far as possible, all statements of fact are supported by original source material including books, newspapers, magazine articles, personal papers and correspondence, and field notes. Quotation marks indicate actual text taken directly from those sources. Certain situations or settings that have not been recorded by someone with personal knowledge (i.e., from actual contemporaneous field notes) have been imagined by carefully following the best information available.

Library of Congress Cataloging-in-Publication Data

Names: Warner, Chuck (Charles H.), author.
Title: Birds, bones, and beetles : the improbable career and remarkable legacy of University of Kansas naturalist Charles D. Bunker / Chuck Warner.
Description: Lawrence, Kansas : University Press of Kansas, 2019. | Includes bibliographical references and index.
Identifiers: LCCN 2018058629
ISBN 9780700627721 (hardback)
ISBN 9780700627738 (paperback)
ISBN 9780700627745 (ebook)
Subjects: LCSH: Bunker, Charles D. | Naturalists—United States—Biography. | BISAC: BIOGRAPHY & AUTOBIOGRAPHY / Science & Technology. | SCIENCE / Paleontology. | SCIENCE / Life Sciences / Zoology / Ornithology.
Classification: LCC QH31.B87 W37 2019 | DDC 508.092 [B]—dc23
LC record available at https://lccn.loc.gov/2018058629.

British Library Cataloguing-in-Publication Data is available.

Printed in the United States of America

10 9 8 7 6 5 4 3 2

The paper used in this publication is recycled and contains 30 percent postconsumer waste. It is acid free and meets the minimum requirements of the American National Standard for Permanence of Paper for Printed Library Materials Z39.48-1992.

Contents

Acknowledgments

A hearty thanks to Bob Timm, curator emeritus, mammalogy, research affiliate, University of Kansas (KU) Biodiversity Institute, research associate for the Kansas Biological Survey, and professor emeritus of ecology and evolutionary biology at KU, who opened the door to the life of my grandfather Charles D. Bunker, and encouraged me to memorialize that early pioneer of the natural history museum. Also, thanks to the late Larry Martin, the always infectiously exuberant paleontologist who possessed the good sense to introduce me to Bob because "Bunker was a mammal guy."

Thanks to my daughter Stacey Hatton for declining the opportunity to write this book. With a blossoming writing career, she initially seemed the likely candidate to take on Bob Timm's challenge. When she announced that prior commitments to family and writing precluded her from taking on such a daunting project, she threw me under the bus by declaring that I should attempt it. Despite my having no experience writing something as lengthy as a book, she insisted that I "could tell a story," and a book was simply stringing together a bunch of stories. Her advice always remained in the back of my mind as I worked on this project.

Thanks to my son, Mike Warner, for providing IT support through this writing odyssey by keeping my laptop(s) up and going, and for listening attentively to my stories about what I had found during my research.

Although primary research documents for Charles Bunker were scant and my skill at digging up secondary resources took years for me to develop, I want to acknowledge the helpful staffs of the following organizations who helped me put together Bunker's story: the Watkins Museum of History, Lawrence, Kansas; the Local History Room at Lawrence Public Library; the Register of Deeds of Douglas County, Kansas; the Fort Wallace Museum in Wallace, Kansas; and the Mendota Museum and Historical Society. Their assistance allowed me to visualize where my grandfather lived and worked through historic maps, old city directories, and census rolls. However, most of the research information came from the University of Kansas Spencer Research Library. For years I have pestered the staff of the Spencer, specifically Kathy Lafferty and Letha Johnson. I thank them for their patience and willingness to coach me on researching.

Through the years I was fortunate to have a wide variety of readers of my manuscript. My first reader was my brother Jim Warner, who is old enough to have an actual memory of Grandpa Bunker. Jim also contributed some distinct personal insight about the man. Incidentally, Jim can still fit into our grandfather's bright red holiday vest, which he still proudly wears every Christmas Day. Other readers include Bannus Hudson and Dave Johnson, longtime friends; Charles "Chip" Jones, a retired journalist and author; Nancy Hause, a retired journalism professor; Jennifer Sanner, editor of *Kansas Alumni*, the magazine of the Kansas University Alumni Association; and finally, Jerry Masinton, professor emeritus and former associate chair of the English Department at the University of Kansas. All read parts or all of my manuscript and offered advice and encouragement, for which I am sincerely grateful.

Readers are not the only ones who deserve acknowledgment. Others offered valuable advice and insight. Early in the process, Lawrence author Bill Sharp, who cowrote *The Dashing Kansan: Lewis Lindsay Dyche: The Amazing Adventures of a Nineteenth-Century Naturalist and Explorer*, graciously shared his wealth of knowledge about the museum and his experiences in writing and publishing a historical biography. Because I am not a hunter, I relied on my friend John Hampton, an avid bow hunter, to help me better understand the psyche of hunters. Jonathan Wilke, who today lives on the property next to where Bunker's old cabin was located, steered me to Karl Gridley, the son of the current owner. Karl took the time to show me the humble remains of the cabin that was the focal point of decades of weekend hunts that trained hundreds of young naturalists. Cathy Dwigans, a Lawrence writer with extensive knowledge of the history of the University of Kansas Natural History Museum, and Martha Skeet, the granddaughter of Charles "Pug" Saunders, shared their valuable information, perspectives, and insight. Also, thanks to Clenece Hills, a former schoolteacher, writer, and radio host, who invited me on her radio show several years before the completion of my manuscript. In addition to her encouragement, that public admission that I was working on a book made it difficult for me to walk away from the project.

A heartfelt thanks to the faculty and staff of the University of Kansas

Natural History Museum. David Burnham, professor of geology and vertebrate paleontology preparator at the University of Kansas, advised me on the proper scientific nomenclature and the basics of fossil hunting. Bruce Scherting, director of Project Art and the Medical Museum at the University of Iowa Hospitals and Clinics and former director of exhibits at the natural history museum at the University of Kansas, patiently shared his vast knowledge of the history of the *Panorama of North American Mammals.* Lori Schlenker showed me the panorama from behind the landscape and answered numerous questions about the architecture of the original Dyche Hall. Finally, thanks to Kris Krishtalka, today's director of the KU Biodiversity Institute and Natural History Museum, for his support, encouragement, and commiseration regarding the perils of getting published, as well as reviewing the scientific accuracy of my book.

A sincere thanks to the staff at the University Press of Kansas. Kim Hogeland, my acquiring editor, was patient and supportive, and an enjoyable collaborator. She gave me good advice and a much appreciated second chance. Likewise, Kelly Chrisman Jacques, my managing editor, shepherded me through the unfamiliar and fascinating world of bringing a book to market, and my copyeditor, Amy Sherman, brought her much appreciated fresh eyes and attention to detail to the project.

Last, but certainly not least, my most heartfelt thanks must go to my wife, Karen Warner. Despite occasionally asking when I thought I would finish the book (the actual research and writing took over six years), she continually offered her support and encouragement. She willingly read and edited the full manuscript dozens of times. More importantly, her advice and counsel contributed enormously to the final product. As a retired elementary teacher who loved to teach creative writing, she homeschooled me on sentence structure, action verbs, and the use of adjectives and adverbs to convey emotion. Karen, you are the reason why I was able to finish this book, and your belief in me is why I have loved you for over fifty years.

Prologue

I don't know with absolute certainty when I actually committed to writing a book about my maternal grandfather, Charles Dean Bunker, but I know exactly when the seed was planted. It happened during a family reunion in September of 2009. Three generations of Bunker's direct descendants were on a behind-the-scenes tour of the museum of natural history on the campus of the University of Kansas. We were all excited to see the place where Grandpa Bunker had worked for most of his adult life. During that visit something unexpected occurred.

Early in our tour we were escorted to the top floor of the historic old building, a part of Dyche Hall not open to the general public and accessible only by key. Our guide, the curator of mammals, showed us the laboratories where today's naturalists and students prepare animal specimens for the collection and conduct scientific experiments. At the north end of the floor was a large workroom filled with an assortment of massive white storage cabinets. Our host swung open one broad door of a cabinet to expose stacks of wide, flat drawers. With a dramatic sweeping gesture, he deftly slid out a drawer to expose the contents: row after row of dead animals. He continued to open more drawers, exposing skins from the mammal collection: ground squirrels, chipmunks, bats, and many more. Moving to another area, he showed us more drawers filled with dead birds. We saw not only every species we might imagine but multiple specimens of each. Their dried skins, complete with the head and feet, were all laid side by side, as if they were lined up for the start of some macabre race. Those collection drawers, with their quirky and morbid contents, totally captivated everyone.

Then our tour guide offered to show us something unexpected: my grandfather's old field notes from his early collecting days. To retrieve them, he led me down a dark hallway to an antique metal door that marked the entrance of a walk-in vault. He slowly swung open a heavy door that guarded a small, dimly lit room. Overflowing shelves and filing cabinets filled the eight-foot-square space from floor to ceiling. This was the repository of the historical and very oldest records of the museum. Multiple sets of field notes from dozens of the pioneer collectors from the 1800s, most of them now hardbound, lined the shelves. Stacked on a table were large, oversized, leather-bound catalogs with pages of carefully penned entries that recorded the very earliest specimens added to the museum's collections. Our guide walked directly over to one shelf and began reading the labels on some boxes. There, crammed into the middle of a shelf at eye level, were two small, ordinary fiberboard storage containers, each roughly the size of a cigar box.

We carried the cartons back to the workroom and opened them on a worktable. As we sorted through the contents, we found several top-bound notebooks, small enough to fit in a shirt pocket. These were Bunker's field notes from the early twentieth century. They chronicled his scientific collecting trips to various places: the Oklahoma Territory, western Kansas, and Alaska. In addition, there were several papers he had authored: one defending the English sparrow, a species of bird that a group of farmers wanted to eradicate, and another on how to clean flesh from animal skeletons using insects.

As our tour concluded, the curator suggested that I consider writing a book about Bunker's life. To encourage me, he offered to make the contents of the old cartons available for review in more detail. Although I had no previous experience writing something as substantial as a book, I agreed to borrow the field notes and papers. Whether I would take up the challenge of writing a book hinged on whether I had the time, the interest, or the talent for such an endeavor. Since I had recently retired, time was not an issue, but during my business career I had never written anything longer than a business letter. Also, I wasn't sure if a book was merited, since I knew nothing of Charles Bunker's professional contributions; to me he was just my grandpa who worked at the natural history museum. In any event, without making any firm decision as to whether I would commit to writing a book, I left the museum with some field notes and documents from the old museum boxes.

A few days after the family reunion, I began by diving into Bunker's dog-eared field notes, handwritten on yellowing pages. Deciphering his handwriting and his careless attention to grammar and spelling proved to be a considerable challenge. I soon discovered that transcribing his notes to the computer improved my comprehension. With the help of that exercise, a complete, candid, and unabridged picture began to emerge of my grandfather as a person, as well as an authentic description of the hard life of collecting trips, especially in the early twentieth century. Those documents provided me with a historical treasure trove that opened a door to the past with my grandfather as the guide.

Despite my fascination with his field notes, I soon realized that I would need additional information to tell the complete story of Charles Bunker. As a supplement to the materials from the vault in Dyche Hall, I sought out genealogy records; newspaper articles; city directories; books about the history of Lawrence, Kansas, the university, taxidermy, and natural history museums; and anything else that would tell me more about this man I never really knew (he died when I was just two years old). I also discovered the research library at the university, the local historical society, the public library, and the county's register of deeds. I soon concluded that the written and recorded history of Charles Bunker was sketchy at best. Needless to say, that made sense considering how he always tried to avoid the limelight. The months turned into years while I continued my research. Fortunately, I discovered that I enjoyed the process of peeling away rings of the onion of his life. Every new tidbit of his history gave me a great sense of satisfaction and accomplishment.

As I researched, it occurred to me that Charles Bunker was not the typical history book hero who led a country, won a battle, or discovered a cure for a dreaded disease. On the contrary, he was the kind of man who could be easily overlooked in a crowded room. On those rare occasions when he attended large gatherings, he intuitively gravitated toward the sidelines to avoid attention. Likewise, whenever he was part of a group photograph, his shyness invariably drove him to the back row. His medium height, trim build, and lack of any outstanding physical features also contributed to his anonymity. Add to that his enjoyment of working with his hands and always living a frugal life, and it was no stretch of the imagination for people who did not know him to consider him ordinary. So why is his story worth saving? First of all, appearances can be deceiving — his accomplishments were far from commonplace. Secondly, his story illustrates how those who lead quiet lives can still make a significant difference in the world around them, a message that seems important in this world of status and wealth worshipping.

Throughout most of my time researching, another concern continued to nag me: Was I too close to the subject because he was my grandfather? Would a book about Charles Bunker be questioned because it lacked objectivity? To overcome that reservation, I needed independent validation of the importance of his life and career, and assurance that his story would have value and resonate with people in general. I needed something more than a memory of my mom saying how she thought her dad had "hung the moon." Eventually, something occurred that reassured me that his story stood on its own merits and would not be dismissed as personal family bias. During the research phase, I took the opportunity to meet with today's faculty and staff at the museum. In the course of those encounters, when I told them I was researching Charles Bunker, everyone knew of him and most could easily recall many of his scientific and curatorial contributions. I thought that was unusual because he had been gone for seventy years. In fact, no one at the museum today is old enough to have actually met him. But more significantly, when I mentioned I was Bunker's grandson, their faces, every one of them, lit up like Christmas morning. The sincerity of their reactions told me that my grandfather's life was special and his career was entrenched in the history of the museum; his story was part and parcel of its fabric and lore. Although my introduction to the field notes and papers served as a seed for this project, in the end, it was the accumulation of those authentic and innocent reactions from today's faculty and staff that ultimately convinced me to preserve his story.

One point of clarity: the original name of the institution where Bunker worked was the State University of Kansas. Over the years various names have been used interchangeably, including the University of Kansas (today's official name), Kansas University, or the short form of simply KU or KSU. When the Kansas State Agricultural College in Manhattan, Kansas, officially changed its name to Kansas State University in 1959, KU discontinued using KSU and State University of Kansas to avoid any problem distinguishing between the two schools. Looking back, losing the name State University of Kansas was probably for the best, considering that it created the unseemly acronym "SUK."

Birds, Bones, and Beetles

CHAPTER ONE

Migrating to Lawrence

With the sky's first glow on a wintry morning in March of 1891, a young man in his early twenties and his parents stood shivering on the train station platform in Mendota, Illinois, waiting for the westbound train. Like most men of the day, father and son each wore a suit and tie, plus an overcoat, muffler, and newsboy-style cap to keep warm. Dad was in his early sixties, five foot six, with black hair and a dark complexion; his son was slightly taller and had brown hair, narrow shoulders, and fair skin. Mom, a pleasant-looking woman with an average build, was in her late fifties and wore a long dress, warm coat, and winter bonnet. As a working-class family they couldn't afford stylish clothing, but what they wore was certainly suitable for church.

This journey marked a momentous event for the family. Pulling up roots from the place they had called home for decades to relocate to an entirely unfamiliar town several states away filled them all with apprehension and excitement simultaneously. To make matters even more unsettling, they were able to bring only their most essential belongings. Most of their clothing occupied a handful of satchels and suitcases that they planned to carry onto the train. A single steamer trunk stuffed with family mementos, some household items, and their hunting rifles and shotguns rested next to them on the platform. The rest of their possessions and furnishings either had been sold to help pay for their relocation or left behind in Mendota.

As soon as the westbound train pulled in to the Mendota train station, it came to an abrupt halt. Knowing that they had only a few minutes to board before the train would depart, the father said to his son, "Charles, help your mother with the bags while I make sure our trunk is loaded onto the baggage car." The son's full name was Charles Dean Bunker. He had been named after his uncle, Charles Dean, who had married his father's sister Rachel. However, nobody called him by his full name except for his parents, David and Susan Bunker, and then only if he was in trouble. Throughout his childhood, his parents and the people of Mendota called him simply Charles — a fine name for a serious and quiet boy.

David Bunker quickly climbed the stairs to the passenger car to join his wife and son in the coach compartment. Many of the seats were occupied, but they found an empty row with seats facing each other. Given the warmth of the passenger car, they began to thaw from their time on the train platform. Then they stowed their winter coats with their luggage above and below their seats. As they all settled in for their daylong train ride, Charles took his place by the window. After a few minutes the conductor signaled to the engineer and the train began to slowly pull away from the station.

A quarter of a mile down the line, the slow-moving train crossed the bridge over Mendota Creek, where Charles spotted their old house on the edge of town. As it was the only home he had ever known, he was already feeling a sense of loss. Mendota was a good place to grow up. The townspeople worked hard, went to church, and raised families. Everyone knew their neighbors and each other's business, and generally looked after one another. Taking his one last chance, his eyes scoured the familiar fields that surrounded the town. He had wonderful boyhood memories of hunting in the woods and wading in the gurgling stream near his childhood home. As the train picked up speed, Charles sighed as he saw Mendota growing smaller in the distance.

Departing Mendota by train seemed both appropriate and symbolic because of how the western expansion of the railroad played such a pivotal role in Mendota's history. Illinois was granted statehood in 1818, when the prairies of Illinois were inhabited primarily by Native Americans. Three decades later, around 1850, pioneers gradually began to settle and farm the fertile soils of central Illinois. At that time Chicago had not yet reached the status of a major city, and only a few small communities and trading posts dotted the state. Galena, Illinois, stood out as a shining economic beacon in the state. Located in the far northwest corner of Illinois, its lead mining activity was in full swing. Initially the mines shipped ore down the nearby Mississippi River on steamboats. But because of the difficulty of navigating that portion of the river, a group of investors sought to build a railroad. In 1851 the state chartered the Illinois Central to construct a railroad from East Dubuque, just west of Galena, to Cairo, at the southern tip of the state at the confluence of the Ohio and Mississippi Rivers. The construction of the road bed began in 1852. At the same time, another infant railroad venture, the

Chicago and Aurora, a forerunner of the Chicago, Burlington and Quincy, was pushing westward from Chicago. By November of 1853 the tracks of the Chicago and Aurora met those of the Illinois Central ninety miles west of Chicago, establishing a station that was named Mendota. According to the *Magnificent Whistle Stop: The 100 Year Story of Mendota, Ill.*, published by the Mendota Centennial Committee in 1953, "the station was given an Indian name meaning 'where two trails meet.'"

Because there were not enough workers in Illinois to construct the 705 miles of track, the railroads recruited men from back east and Europe. That resulted in a rapid influx of migrants to Mendota that included Irish, English, German, and Danish people. To feed and house these newly arriving workers, Mendota experienced a sudden building boom. In 1854, when newlyweds David and Susan Bunker learned of the need for carpenters in Mendota, they immediately moved there to settle into their married life.

David and Susan had first become acquainted in Michigan a few years earlier. David Bunker was born in 1830 in Cardington, Ohio, to Slocum and Matilda Bunker. Several years later he and his family moved to Marshall, Michigan, to pursue new business opportunities. David was the second child of seven and the only son. His father left his family in Marshall, making his way north about fifty miles to what would become known as Hastings, Michigan. There he became involved in the lumber industry. By 1838 the family joined Slocum in Hastings, where he had successfully established a saw mill, a trading post, a post office, a tavern, a school, and a grist mill.

Born in Michigan in 1833, Susan Jane Spencer lost both of her parents by the time she was seven. She was taken in by a neighbor who unfortunately took advantage of her by making her toil as her personal drudge. Eventually Susan ran away and stayed a short time with an Irish family before a good friend of her parents took her in and raised her. Ironically, that Dickensian coming-of-age experience contrasted with the life of her maternal grandparents, who belonged to the British aristocracy. Lord and Lady Crosby from Sheffield, England, had disowned Susan's mother when she married Susan's father because he was a simple commoner.

David's younger sister Rachel was the first of the Bunker family to meet Susan, while the two girls were attending the same boarding school. During one Christmas vacation, Rachel invited her classmate to spend the holiday with the Bunker family, and there Susan first met David. After that initial meeting, the two courted sporadically for several years while she worked as a governess for the governor of Michigan and David attended college in

Ann Arbor, Michigan. When Susan received a letter from David proposing marriage, she immediately accepted. Contrary to his family's plans, David dropped out of college and the two married on May 14, 1854. His father had hoped that David would take over the family business, but such a life held no interest for David. It was time to strike out on his own enterprise.

Upon their arrival in Mendota, David and Susan Bunker were amazed when they observed the fruits of the early stages of the building boom. In just the previous year the town had seen the completion of a grocery store, drugstore, brewery, public school, and two meat markets. Under construction was an impressive stone structure to house a steam-powered flour mill, which would prove vital to the farming community. Not only did David find work as a carpenter, he and Susan soon learned that the small midwestern town was a good place to live, work, and raise a family. The Bunker family eventually moved to a small house in the 200 block of south Twelfth Street, near the edge of town, within a block of the Chicago and Aurora westbound rail line and adjacent to a small stream that ran through town, called Mendota Creek.

As David and Susan assimilated into the social fabric of Mendota, they started their family. Their first child, a son named Louis, was born in 1855, followed by William in 1857 (who subsequently died in 1860), and Caroline Elizabeth (Lizzie) came along in 1859. According to the 1860 US Census, David continued to work as a carpenter. Their next child, a son named Harry, arrived in 1862. That was also the year that David enlisted to fight for the Union in the Civil War. At thirty-two years of age, he signed up to join Company C of the 104th Regiment of the Illinois Infantry. His first combat was the Battle of Chickamauga, in Georgia, on September 19–20, 1863. During the battle David was seriously wounded by a shot in the abdomen. Initially hospitalized in Nashville, he was later transferred to a hospital in Louisville. After recuperation, he rejoined his regiment in February 1864 and was promoted to first sergeant. The regiment was returned to Chattanooga, Tennessee, so they could prepare to join in the Union assault on Atlanta. On May 14, 1864, his wedding anniversary, David was wounded a second time at the Battle of Resaca, just north of Atlanta. Again, he was hospitalized in Louisville. Upon his recovery, he remained at the Union hospital as a ward master, tending to the needs of the recuperating patients. He served there until he received his honorable discharge on June 12, 1865, in Louisville.

When David returned from the war, he found Mendota bustling with the population approaching three thousand. The town now boasted nine dry goods stores, sixteen groceries, four hardware and tin stores, three furniture stores, two steam grain elevators, two flour mills, one foundry and machine shop, one plow factory, one planing mill, three hotels, and two banking houses. In addition, there were seven churches of different denominations, six public schools, and two colleges. David soon found work as a wagon maker, and he and Susan began again to increase the family. Julie and Alice arrived in 1866 and 1868, respectively, and Charles brought up the rear in 1870. With Charles' birth, the Bunker house overflowed with children.

By the time Charles was born, Mendota had grown to around five thousand people. The railroad was well established, with nearly forty passenger trains per day traveling through Mendota. Although the construction of the railroad had driven the economy of early Mendota, by the 1870s the town began to rely on being a commercial center for the surrounding farming community. The town developed a reputation as a "Saturday Night Town," where every Saturday afternoon hordes of farmers came to get their shopping done, visit with friends, and generally socialize the day away. In addition, a wide variety of small manufacturing companies had sprung up and provided good-paying jobs for the immigrant railroad worker after the initial flurry of railroad construction. Those businesses included companies such as Mother Hubbard, which produced washboards; the Extra Pale Brewery; Schaller and Goebel, a maker of buggies and carriages; and Tewksbury & Carpenter, which manufactured musical organs.

With a carpenter/woodworker as the primary source of income for the household, the Bunkers never enjoyed much money. Regardless, they always had food on the table, even if the meat mostly came from inexpensive cuts. As for clothing, there were lots of hand-me-downs, especially for Charles, as the baby of the family. Charles learned to appreciate working with his hands by watching his father work with wood. He also understood what it was like to live in modest surroundings, where the family lived from one payday to the next. As a result, Charles learned at an early age that if he wanted something, he had to work for it. Hard, physical work was part of life, and efficiency was something to be valued and respected. Even though the Bunker family fit right into the social fabric of their community, their way of life could best be summarized as only "fair to middlin'."

For the early citizens of Mendota, education topped the list of what they desired for their young community. The first grammar school opened in 1854 on the west side of town, the same year that David and Susan arrived. The next year a second school was established on the east side of town because the residents felt that crossing the railroad tracks to attend school was too dangerous. In the 1860s several colleges were established by various local churches and the West Side High School graduated its first class in 1876.

Although Charles was bright and inquisitive, and schooling was readily available, frequent illness stifled his formal education. There is no specific record of his exact ailment, but the medical recommendation was "fresh air." Following that advice, David and Susan encouraged Charles to spend much of his time outdoors. A boy who instinctively took pleasure from nature, he welcomed the opportunity to spend hours on end exploring the fields and woods around Mendota. As a young boy he constantly collected turtles, bird eggs, and anything else his parents would let him keep. Eventually he learned to identify all the species of the birds, mammals, and reptiles that lived around Mendota. With his inquiring mind, his self-taught knowledge of nature grew, probably more than it would have if he had been in the classroom. In the end, his sporadic school attendance delayed his graduating from grammar school, which most people take to be the cause for why he cut short his education. But his shyness and stage fright made reciting unbearable, which also likely contributed to his ending his formal schooling after the eighth grade.

When Charles was old enough to handle a gun, his father taught him how to shoot a rifle. Eventually he became an accomplished marksman and often, with considerable pride, brought home a squirrel or bird for supper. It was his shooting of game, especially game not destined for the dining table, that eventually introduced Charles to the fascinating world of taxidermy. The local taxidermist was a farmer with the last name of Banning, who mounted local animals as a side job. The first time Charles visited Mr. Banning, he knocked cautiously on the door of the farmhouse. Mrs. Banning directed him to a one-story wood frame outbuilding. Charles knocked guardedly on the door of the workshop. After a moment, Mr. Banning suddenly swung the door wide open. Taken aback, Charles stuttered to explain his interest in seeing the farmer's stuffed animals. As Banning considered the request, he chuckled and admitted that he never had many visitors. Eventually he gestured for Charles to come in. As soon as he walked through the door, it became immediately obvious why Banning had so few visitors — the odors of decaying flesh and pungent chemicals immediately greeted Charles. But if there was a moment when he questioned his decision to visit the taxidermist, his curiosity quickly trumped the stench.

As Banning led Charles toward the back of the shop, the young boy's eyes began to adjust to the darkness of the workroom. Now he could see that above him on high shelves surrounding the room were rows of mounted birds. Charles was fascinated by the eerie specimens of owls, hawks, and crows that seemed to scowl down at him from the shadows. Mr. Banning explained that he specialized in mounting birds, primarily for sale to saloons. Considering how plentiful drinking establishments were in railroad towns, Mr. Banning admitted it was a profitable occupation.

A skylight brightly illuminated Mr. Banning's workbench, which dominated the rear of the workshop. Center stage, a sizable disemboweled crow lay unceremoniously across the stained work surface. Next to the bird Charles noticed a field guide to indigenous birds. Lying alongside were variously sized knives, tools, and needles with thread that Banning used to mount his specimens. Charles and Mr. Banning talked for a short time, the taxidermist fielding a myriad of questions from his inquisitive visitor and opining that taxidermy has always connected man and nature. He also reiterated that being a taxidermist was a lonely existence because not many people were comfortable with the smells and lurid images of dead animals in a taxidermist's workshop. Charles, on the other hand, found the workshop both mysterious and intriguing.

After a while, Mr. Banning looked at his pocket watch, announced that he had work to do, and told Charles it was time to go. Banning had enjoyed their visit, so he invited Charles to come again. Charles would return on a regular basis. Eventually he decided that he wanted to learn taxidermy because of its connection to nature. The idea of working in solitude was a bonus. Once he came to that decision, Charles asked the taxidermist to teach him how to mount birds. Banning thought for a moment and told the youngster that he would do it, but it would cost him ten dollars (over $250 today). Disheartened, Charles informed the taxidermist that he did not have that kind of money. In the end Banning allowed Charles to watch the mounting process on any bird or animal that he brought to the taxidermist. Through this agreement, Banning got new specimens and Charles gained the knowledge he desired.

Charles' interest in birds continued to grow through his teen years and led him and a friend to cowrite an article for a small local ornithology (the branch of zoology that deals with birds) publication called the *Stormy Petrel.* Although the periodical was not highly regarded by scholars at the time, the fact of his article's publication nevertheless illustrates Charles' commitment and enthusiasm for nature and the science of birds at a young age.

By 1889 David Bunker was working as a cabinetmaker for Tewksbury & Carpenter, the manufacturer of musical pump organs. One day he came home and announced that the company had fallen on hard times and was closing. David, who was sixty years old, set about looking for another job and found none in Mendota. Even though the Bunkers had deep roots in their small community, David was forced to look for work elsewhere. Eventually, in early 1891, he found and accepted a position with Haskell Institute in Lawrence, Kansas. Established in 1884 by the United States Bureau of Indian Affairs to assimilate Native American students into white society, Haskell was a boarding school for Indian children and teenagers from various tribes all over the United States. The school taught sewing, cooking, carpentry, and other domestic and industrial skills. Using the skills he previously learned from the wagon manufacturer, David Bunker was hired to teach the students wagon making. By the time David and Susan Bunker were ready to move to Kansas, Charles had no job to support himself living on his own in Mendota, so he elected to move west with his parents.

On their daylong passenger train ride across the Midwest, the scenery was all too similar: field after field of winter stubble, trees lining streams and creeks, all surrounding one small town after another. After a while Charles found the repetitive, gentle, rhythmic bump and sway of the train relaxing. To pass the time, he fogged up the cold glass of the train window with his warm breath. After he wiped it clean with his sleeve, he watched with fascination the tiny cinders and swirling smoke from the locomotive that flew by his window. Despite the chill of the air temperature outside, Charles and his parents remained quite comfortable in the steam-heated passenger car. After their dinner—taken out of Susan's large carpet bag—the train crossed the Mississippi River at the southeast corner of Iowa and began its charge through Missouri. Charles spent most of the afternoon seated by the train window. Most people find the cities and towns more visually interesting, but Charles preferred watching the winter landscape of the countryside. The fallow umber fields and the leafless trees exposed the bountiful birdlife. Even as the train sped along, he could easily scan the fields and recognize his feathered friends. In the Midwest, even in the dead of winter, birds were everywhere, and identifying some of his favorites helped Charles pass the time on his long trip. He made notes in a small pocket notebook about their variety and numbers. Most of the bird species he observed were local, as it was still too early for the migrating birds to arrive. Every few miles, large flocks of birds suddenly took flight when the train rumbled past. They always arose and swirled in unison, as if guided by a single mind. The quickness of their reaction time, which kept them in such a tight formation, never ceased to amaze and fascinate Charles. In addition, he noticed pairs of cardinals, male and female, flitting in the bare tree branches, and solitary hawks perched on the tops of telegraph poles scanning the ground below for an unsuspecting field mouse.

The afternoon shadows were growing longer when the Bunker's train arrived in Kansas City. Fortunately, it was only thirty-five more miles to Lawrence. The Bunkers were excited to see the town that would be their new home, and they were pleased to think that they might be arriving while it was still light outside. As the westbound train from Chicago slowly approached the Lawrence station, it huffed and hissed, finally letting out one long gasp as it came to a standstill. After a minute or so, Charles disembarked the train with his parents. He helped his mother gather their family possessions while his father arranged for transportation. After a few minutes the train finished reloading and was off again on its westward journey.

Once the train was out of sight, silence returned to the Lawrence railway station, a stately Victorian two-story brick structure with white trimmed windows and shutters. Long shadows fell across the winter landscape that displayed patches of snow from a recent thaw. As Charles felt the air temperature drop along with the sun, he observed the glow of lights in the windows of the neighboring houses. In the leafless trees birds chirped and flitted around searching for the last morsels of food before nightfall.

As young Charles, hands in his coat pockets, stood shivering in the cold nightfall, he missed the warmth of the passenger train. Finally, the hired driver with a horse-drawn wagon arrived to carry Charles and his parents to a nearby rooming house for the night. The next day they would look for a place to rent, hopefully near Haskell Institute. After they loaded the wagon they were driven seven blocks from the train depot to the center of town. When they turned south onto Massachusetts Street, they observed a thriving business district. Attractive two-story buildings with brick fronts, each with its own unique architectural style, lined the main street of their new town. The shops and businesses had already closed for the day, but electric street lamps illuminated the four-block-long downtown. All in all, downtown Lawrence impressed Charles.

Behind those buildings, the residential neighborhoods extended out on a grid of streets for eight to twelve blocks to the east, south, and west. The streets were graded, but lacked pavement, which worked well enough for both horse and wagon until they became muddy and treacherous during either a rainy autumn or the spring thaw. For the pedestrians, wooden sidewalks existed only in front of the stores in the business district. As their wagon rumbled past the downtown, the Bunkers came to a beautiful little city park. Leafless full-grown trees and grass trimmed from the previous fall surrounded a small round elevated bandstand.

Despite the genteel appearance of South Park in 1891, Lawrence had not always been so modern and refined. The advent of running water delivered in underground pipes from the Kansas (also known as the Kaw) River, though untreated, had only come about in the last decade. Around the same time, the newly installed sanitary sewer system replaced outhouses. It had only been a few years since the installation of those electric street lights downtown, and both electricity and gas manufactured from coal now illuminated all homes and businesses in Lawrence. By 1891 coal furnaces had replaced wood-burning stoves for heating homes, and it had only been three years since cows and horses had been banned from grazing in this picturesque park.

A block south of the park, the horse-drawn wagon pulled up in front of a well-kept two-story rooming house. The Bunkers carried their bags into the house, storing the large trunk on the covered front porch. They were famished after their daylong journey, so the landlady reheated supper for them. When the family went to bed that night, they all wondered if Lawrence would ever feel as much like home as Mendota.

When the Bunker family arrived in Lawrence in 1891, Kansas had been a state for only thirty years. At ten thousand people, the population of Lawrence was five times greater than Mendota's two thousand. But they had similarities as well: both communities were distinctively midwestern, served as the commercial hubs for surrounding farms, and were home to numerous small manufacturing companies that pumped up the local economy. Education, particularly higher education, was important to both towns, although Lawrence had the distinction of being home to the state university as well as one of the few schools for Native Americans in the country.

In 1891 most adults living in the United States still vividly remembered the pain and suffering of the Civil War, so old allegiances still counted. The fact that Kansas and Illinois had fought on the same side pleased the Bunkers. However, Kansas, and Lawrence in particular, had a much more volatile history. Known as "Bleeding Kansas," their new home state had been at the very epicenter of the national debate on slavery in the years leading up to the Civil War. When the citizens of the Kansas Territory asked to join the Union as a state, US Congress required the territory to decide by ballot if it wanted to be a free or slave state — that is, whether it would prohibit or allow slavery by law. Up until that time the existing states were evenly divided on the slavery issue, so the outcome of the Kansas vote would serve as a tiebreaker and most likely would decide the matter for the entire country. Both the abolitionists and proslavery factions attempted to influence the outcome of a pending election. The slave states, including the neighboring state of Missouri, encouraged proslavery citizens to move to Kansas. To counter that action, staunch abolitionists established the Massachusetts Emigrant Aid Company in Boston, with a mission to finance antislavery supporters to move to the Kansas Territory to settle it as a free state. The efforts of that organization established the town of Lawrence in the Kansas Territory in 1854. After years of political fighting, which even included bloodshed, thousands of proslavery citizens from Missouri crossed the state line to steal a Kansas election by casting illegal ballots. Eventually Kansas voted again, without those election irregularities, to become a free state and was admitted to the Union as the thirty-fourth state on January 29, 1861.

Despite the antislavery vote, the matter was far from settled. As the years of the Civil War dragged on in other parts of the country, bloody skirmishes occasionally flared up along both sides of the Kansas-Missouri border. Bands of guerrillas from both states repeatedly ventured across their border to attack, loot, and kill the citizens of the other. Kansans referred to the marauders from Missouri as "border ruffians," and Missourians called the Kansans "jayhawkers." After each raid, the aggressors would scurry back across the state line to home and safety. Unfortunately, each new attack caused the other side to retaliate with a raid of greater brutality. The local history of both states cast the blame for the bloody raids squarely on the other state, but neither Kansas nor Missouri was blameless. Those repeated raids by both sides resulted in an intense rivalry and hard feelings that still exist today between the two states, albeit without the bloodshed.

The culmination of those raids on Lawrence occurred in 1863 and proved to be the most tragic of all the border raids. In the early morning of August 21, William Quantrill and his raiders from Missouri viciously attacked

Lawrence. By the time they made their hasty retreat, they had killed nearly two hundred men and boys and burned most buildings in the town to the ground. In an odd twist of fate, at least as far as the Bunker family history, Quantrill had lived in Mendota shortly before the war, where by some accounts he taught school, though other reports said he worked loading lumber.

CHAPTER TWO

Aspiring Taxidermist

The morning after their arrival in Lawrence, the Bunkers began getting acquainted with their new hometown. They all left the boarding house early to look for a place to rent. In the 1700 block of Massachusetts Street they found a two-story frame house that had a living room and kitchen on the first floor, a bath and two bedrooms on the second, and was partially furnished. In addition, it was conveniently located across the street from a grocery store and less than a mile from Haskell Institute. David Bunker struck a deal with the landlord and the family moved in.

The following morning, David left the house early for his first day of work at Haskell. After clearing the breakfast dishes, Susan and Charles walked across the street to visit the grocery store. In 1891, according to the city directory, approximately forty small neighborhood grocers were spread out all over the town of Lawrence. With a population of 10,000 citizens, that equates to one store for every 250 residents, allowing most families to walk less than three or four blocks to buy groceries. Typically, the stores were small, maybe twenty-five feet wide and fifty feet deep, offered credit until payday, and served as the neighborhood gossip center. The grocer, Scott Holloway, warmly greeted them and helped them with their shopping.

Although Charles understood that he needed to look for a paying job, he wanted to explore the town first. After he carried the groceries home for his mother, Charles decided to see where his father was working. As he quickly walked past what is known today as 19th Street, the town promptly transitioned to countryside with farms and pastures. It was a sunny day in northeast Kansas, where March weather is either an extension of winter or the beginning of spring. That day it was winter. Charles had covered his short brown hair with his wool cap with a bill and pulled the collar up on his hunting jacket, but still the cold air stung his face.

As Charles neared Haskell Institute, he could see that the campus included a number of large multistoried limestone buildings. He was impressed that his father was working at what appeared to be such a prestigious institution. He stopped and stood there in the cold, reconsidering his plan to visit his father on the first day of his new job. Charles knew that if he saw his father, he would be introduced to strangers, which would require making polite conversation. Such situations always made him uncomfortable, so he turned around and started back toward town.

As Charles headed back along Massachusetts Street toward home, his eyes were drawn to several massive, multistory buildings clustered on top of a large hill off to the northwest. Nearly two miles away and across the treeless prairie, the buildings dominated the skyline, presiding over the valley like a feudal compound. Today that view is nonexistent at street level because of all the trees in the older neighborhoods of Lawrence. Curiosity got the best of Charles, so he decided to get a closer look. He followed a frozen dirt road that led him up a steep incline on the south side of the hill. Once he arrived at the top, he paused to catch his breath. There he observed three large limestone buildings, each several stories high. Dozens of people, mostly young adults, scurried in and out of the largest building. The men wore coats, ties, and hats, and the women were dressed in long skirts and bonnets; they all seemed to stride with great intention and confidence. He wondered what went on inside those buildings, but he was too shy to approach anyone to ask.

For several minutes Charles took in the view from the crest of the hill on the east lawn of the largest building. Unprotected from the bitter wind, he stood shivering. His vantage point gave him a spectacular view with a beauty that caught him off guard. Able to see nearly ten miles in three different directions, he looked to the north across the Kansas River Valley, south across the Wakarusa Valley, and east to where the two rivers converged. That grand panoramic view made quite an impression on the young man enamored with the outdoors and nature. After being transfixed by the view for a few more minutes, he walked back down the hill to the warmth of his new home, still uncertain what institution occupied that beautiful setting.

Although Charles was living at home and his father was gainfully employed, he still needed to find a job, especially if he ever wanted to live on his own. Thinking he would like to find work in taxidermy, the next day he stopped by Holloway's Grocery to ask if there was a taxidermist in town. Charles was glad when Scott Holloway cheerfully greeted him as he entered the store. After the two men had conversed for a few minutes, Charles told the grocer that he had worked with a taxidermist back in Mendota and asked if there was anyone around Lawrence mounting animals. To the best of Holloway's knowledge, the nearest taxidermy activity was at the State University of Kansas. He recalled that the museum of natural history was mounting some very large animals, like buffalo and moose. Initially Charles had only envisioned himself working on birds and maybe some small mammals, but the prospect of working on large animals sparked his interest. He asked for directions to the university and the museum and immediately realized that he had been there the previous day; it was the place with the extraordinary view. Holloway also suggested that Charles ask for Professor Dyche, as the person in charge.

That pleasant exchange between Charles and Holloway, and the grocer's advice, would eventually alter the direction of Charles' life. Charles walked five blocks north before turning left on Adams Street, today known as 14th Street. The steepness of the dirt road leading to the main entrance to the university reminded him of stairs. Once again, when Charles reached the campus at the top of the hill he paused to catch his breath. Fraser Hall, the largest building on campus, towered before him. To the west of Old Fraser stood Snow Hall, the home of the KU Museum of Natural History, a substantial three-story limestone building on the westernmost edge of the campus. Although Snow Hall was not as imposing as Fraser Hall, that distinction did little to alleviate his uneasiness about asking for a job. Pacing back and forth next to the front steps of Snow Hall, Charles could not imagine what he would say to someone as educated as a professor. Anxiety overcame him. He knew so little about the museum, much less the university — he worried that the taxidermy work here might be very different from his experiences in Mendota. Despite his misgivings, he was finally able to gather himself enough to enter Snow Hall to find Professor Lewis Lindsay Dyche.

As soon as Charles opened the front door of Snow Hall, those familiar noxious odors hit him, reminding him of Mr. Banning's modest shop back in Mendota. He followed his nose to the museum workshop on the third floor of the building. The stench, usually enough to keep most people away, only served to entice young Charles to enter. As he peered through the open door, the entire workroom appeared as busy as a beehive, immediately shattering any sense of calm he may have previously attained. Nearly a dozen workers and students scurried about, tending to assorted animal specimens spread out on large worktables in various stages of deconstruction.

Taxidermy, both here and back in Mendota, was basically a centuries-old process to physically replicate the form and position of animals in life by stretching and sewing treated skins over artificial forms. The French first coined the term *taxidermy* in the early 1800s, from the Greek roots *taxis*, "arrangement," and *derma*, "skin." However, the similarities between Dyche's and Banning's workshops ended with the smell and basic process. In the museum workshop, the mounts were considerably larger and more spectacular than Charles had ever seen in Mendota. For instance, partially completed mounts of a large female bison and a male moose (animals that he had never seen before in real life) occupied the center of the room. The shelves and cabinets that surrounded the room displayed skeletons and skulls from dozens of mammals as well as a variety of snakes and frogs preserved in jars of alcohol solutions. The pace of the commotion and the variety of work fascinated Charles. Eventually he would learn that much of the workshop activity that day related to an extraordinary display of large mounted animals slated for the Kansas state exhibit at the upcoming World's Fair in Chicago, officially called the World's Columbian Exposition.

Oh my, he thought — this was certainly not Mr. Banning's taxidermy shop. Charles asked the person nearest the door where he could find Professor Dyche. The person he asked — J. Charles "Pug" Saunders, the senior taxidermist for the museum — directed him to the confident gentleman in the center of the room directing all the activity. Intimidated by Dyche's presence, Charles shyly approached the professor and in a slight voice explained that he was looking for work as a taxidermist. Dyche cut him off short, telling him that all of those positions were filled. Dejected, Charles quietly exited Snow Hall and walked back down the hill. Despite the rejection, witnessing the workshop firsthand convinced him more than ever that a job at the museum was what he wanted most.

Although Dyche's rejection crushed Charles, he set out early the next day looking for work. He understood that given his youth and lack of experience he could not be too picky about what job he took. When the owner of the Hoadley Printing Company offered him an entry-level position as a pressman, he accepted. The print shop was downtown, in the basement of Lawrence National Bank, a large ornate stone bank building on the corner of Winthrop (today called 7th Street) and Massachusetts Street. Every day when he walked to work along Massachusetts, he glanced longingly up the hill toward the university and museum. Even though his new job would allow him to work with his hands and the working conditions were acceptable, he knew he would miss taxidermy and the opportunity of being outside in nature. Regardless of his disappointment, Charles never gave up on his dream of one day working for the university.

In the fall of 1892 Holloway told Charles that he had heard a rumor about Professor Dyche taking a large quantity of mounted animals to Chicago for the World's Fair that was to open in May of 1893. Like most everyone in town, Charles could not imagine the scale and scope of Dyche's endeavor, but he soon discovered, much to his amazement, that the entire menagerie of mounted animals was so large that it would require eight railcars, including one flatcar for the larger animals and one car for the professor and his staff to live in while they were setting up the exhibit. In November the empty railcars arrived and were parked on a rail siding near the Lawrence train station. For several weeks, wagon after wagon of mounted animals, crates, and equipment were ferried down the hill from the museum and loaded onto the cars. Everything about the process fascinated Charles, so whenever he could find a few minutes of spare time, he walked over to watch as the incredible array of mounted wild animals was carefully placed aboard the train.

The 1893 Columbian World's Fair in Chicago was a pivotal moment in Dyche's career. He prepared and curated an exhibit that eventually included 122 mounted North American mammals. To showcase the animals, he built a landscape complete with actual trees and mountains constructed of a wooden frame and papier-mâché to offer an appearance of the animals in their natural setting. Dyche arranged all the animals in groupings, a technique he learned from his old teacher, William Temple Hornaday, from the National Museum in Washington, DC. Dyche recognized that not all museums valued group displays, thinking that it verged on entertainment and pulled focus away from scientific study. However, Dyche believed that putting the mounts in scenes created context about the animals' lives and made the specimens more lifelike. In fact, Dyche's World's Fair exhibit included a scene of a mountain lion standing over the carcass of a freshly killed deer. Dyche did this because he believed the viewer would better understand the sometimes cruel realities of how the animals behaved in nature.

Dyche's menagerie of mounted wild animals took Chicago by storm. Not only was his collection the centerpiece of the Kansas exhibit but many hailed it as one of the high points of the entire fair. An 1893 article in *Scientific American* declared, "In the north wing of the Kansas building is one of the most remarkable exhibits to be seen at the great Fair . . . the work of a man who is recognized by naturalists as the best taxidermist in the country, if not in the world." Estimates reported that crowds visiting the Kansas exhibit exceeded twenty thousand during the final weeks of the fair. Up until that time, most of the nation considered Kansas uncultured and undereducated, a desolate place on the edge of the frontier. Dyche and his exhibit put Kansas on the map and secured his own reputation as a world-renowned naturalist.

Ironically, despite the positive national press, many of the local and regional newspapers criticized Dyche and his exhibit. An article in the *Leavenworth Times* on December 17, 1892, declared, "We have yet to see one paper published in Kansas, outside of Lawrence, that endorses . . . making the stuffed animals the principal part of the Kansas show." The article continued, "A show of our grains, grasses, minerals and other products will be of far more importance to Kansas than this show of the skill of one taxidermist. . . . Very few Kansas people want Kansas known to the world as the stuffed animal state."

The excitement generated by Dyche's exhibit only served to further increase Charles' desire to work at the museum. After Dyche returned from Chicago, Charles occasionally dropped by the museum to see what projects they were working on and to ask if there might be any work for him. He even stopped Dyche on the street one day to ask about a job. Sadly, each time the professor turned him down.

Finally, in the fall of 1895, Charles gathered up his courage once more to call on the now world-famous director of the KU Natural History Museum. It was early October, just after the professor had returned from another very successful hunting expedition. After two months' hunting in the northern reaches of Greenland and the Arctic Circle, he triumphantly returned to the university with a grand collection of polar bears, caribou, seals, and walruses. On that day, Charles found Dyche in the museum workshop of Snow Hall, hands on his hips, chest held high, and now sporting a mop of long brown hair. Again, Dyche's presence intimidated young Charles. He shyly approached the professor with his hat in hand to once again ask for a job. For Charles, such a simple act took great determination, given his lack of confidence and how he had been rebuffed repeatedly. He fumbled to explain his life's ambition of mounting birds. This time the usually impatient professor gave the appearance of actually listening. As Charles rambled on nervously, Dyche quickly calculated in his head how the young man could

1893 Chicago World's Columbian Exposition (University of Kansas Natural History Museum records, University Archives, RG 33/0 Photographs, Kenneth Spencer Research Library, University of Kansas Libraries)

help with all the specimens that had recently been shipped home from the far north. Charles' timing was perfect. After further discussion, the professor made him a job offer, at fifteen dollars per month. Despite the paltry salary, Charles accepted, thinking that at last he had his foot in the door of the museum and the world of taxidermy. Reminiscent of his parents when they were first married, Charles was following his heart; the money didn't matter. Since he was still living at home with his parents, he knew he could get by on the small salary. At last he had a job that would scratch his itch for birds and taxidermy.

At the time, Charles knew very little about Dyche, other than that he was a skilled taxidermist who had wowed the Chicago World's Fair. Fortunately for Charles, he would have the opportunity to work under a highly intelligent and a gifted taxidermist. Dyche had begun his meteoric rise in academia when he came under the tutelage of Francis Huntington Snow, one of the first faculty members of Kansas University and the director of the museum that was housed in the building that was named after him. Despite a late start beginning his studies at the university, Dyche excelled as a student to the point that he taught classes as an undergraduate, graduated with two bachelor's degrees at the top of his class at the age of twenty-seven in 1884, and gave a commencement address. By 1889, worried that an eastern museum might try to hire Dyche away, Snow asked the board of regents to promote Dyche to professor. Finally, after repeated prodding, they appointed Dyche to the position of Professor of Anatomy and Physiology, Taxidermists, and Curator of Mammals and Birds, effective July 1, 1889. Despite his new title, the students had been calling him "professor" for years. With that promotion, Snow retained the title of Director of the Museum of Natural History, but turned much of the day-to-day management of the museum over to Professor Dyche, which included organizing an increasing number of collecting expeditions; in 1889 he went to British Columbia for six months to hunt mountain goats and bighorn sheep, and in 1890 he journeyed to Minnesota for moose and elk. The collection was growing and was poised to take on the upcoming World's Fair.

When Charles arrived at the museum on his first day of work, Dyche surprised him at the door and excitedly announced that they both were taking the train that afternoon to Topeka, where he was to give a lecture that evening. In the late 1800s one of the most popular forms of entertainment and culture involved speakers — including writers, preachers, and experts — who traveled from town to town giving lectures on various subjects. Their compensation came from the admissions charged. Seeing a financial opportunity, Dyche organized a series of lectures to expound on the excitement and adventure of his recent expedition above the Arctic Circle. The professor explained that he needed Charles to operate the magic lantern, which projected photographic images from glass plates. Dyche was a consummate promoter and understood that projecting photographs from the expedition greatly enhanced his speech. Although the unexpected change of plans caught Charles off guard, he was always willing to do whatever was asked.

Late that afternoon the two men exited the train in Topeka and Dyche flagged down a carriage to take them to the theater. Charles labored to carry most of the gear, which included the projector and a bag of Dyche's fur outerwear made and worn by the indigenous people in the Arctic. They arrived just in time to set up the projector and make sure there were enough chairs for the audience. Just before he was to begin his presentation, Dyche

KILLING THE HARPOONED WALRUS.

Lewis Lindsay Dyche in fur suit demonstrating how to harpoon a walrus as he would have appeared at his lectures on his speaking tours in the late 1890s (University of Kansas Natural History Museum records, University Archives, RG 41/0 Photographs, Kenneth Spencer Research Library, University of Kansas Libraries)

changed into the fur coat and pants to show what he had worn while hunting in the Arctic Circle. Dyche was quite warm in the fur outfit by the time the crowd assembled in the auditorium, but wearing it made his demonstration of how to use a harpoon more exciting. The strong voice and theatrical oratory Dyche used to speak to the two hundred people in attendance impressed Charles. The slideshow consisted of dozens of photographs of Dyche hunting, killing, and skinning polar bears and walrus, and thoroughly mesmerized everyone in attendance, including Charles. Several times a week for the next year and a half, the professor put on hundreds of presentations around the state and region, each time with Charles helping quietly in the background.

If Charles had any reservations about whether helping Dyche with his lectures distracted from learning the skills of taxidermy, he soon realized that the benefits outweighed the time away from the workshop. When he first joined the museum, Charles lacked considerable significant technical scientific knowledge, especially of taxonomy — ascribing the proper scientific names for things found in nature. That was no wonder, considering that Banning, his former mentor, lacked formal scientific training. Charles found Dyche to be a fountain of information, but more importantly, someone more than eager to share his knowledge and expertise. Whether traveling on the train to the next lecture or in his speeches, the professor always communicated as a scientist, often using the scientific names of the animals even when it wasn't required. Charles gladly soaked up any technical information Dyche shared, and that newly acquired information, along with his years of observing animals in nature, improved his skills in identifying species, which would prepare him to later work as a qualified custodian of a zoological collection.

Operating the projector for Professor Dyche also gave Charles the unexpected opportunity to learn scientific theory. In addition to sharing his hunting adventures, Dyche usually took the time during his lectures to expound upon various contemporary scientific subjects like Charles Darwin's theory of evolution and his assertions concerning the survival of the fittest. Charles learned best by doing or watching rather than from reading, so for him those lectures by the professor were mind-expanding experiences. He had always loved nature, but this exposure gave him a totally new perspective on the world around him. It was like auditing a college-level science class, something he had never dreamed he would do. As it turned out, all that time with Dyche on the lecture tour would provide much of the necessary groundwork for Charles' career in the museum.

Although the two men shared a strong interest in nature and wildlife, they differed in character, aspiration, and affect. Charles would discover that Dyche, as a confident museum promoter, often curried favor with university leaders and state politicians, constantly focusing on the vision of the museum and looking at the bigger picture. Charles, on the other hand, was always more comfortable with his coworkers and students, working in the background and attending to the material aspects of the collection. Dyche would offer valuable scientific knowledge to Charles; in turn, Charles would take care of things more mundane so Dyche could pursue the larger goals. In the years that followed, this symbiotic relationship between Charles and Dyche would prove valuable to both men as well as to the museum. Their contrasting personalities complemented each other, like a marriage of opposites where the strength of each balanced the other's weakness. And like a good marriage, their relationship would last for years.

When Dyche offered him the job in 1895, the natural history museum was a bit of a mystery to Charles — which is why, by his own admission, he told Dyche that he simply wanted to be a taxidermist who mounted birds. Naturally, Charles considered taxidermy from Mr. Banning's frame of reference, in which preparing a mount for a sportsman or selling a specimen to a bar owner was a simple commercial transaction. However, he would soon discover that the natural history museum at the University of Kansas, like any natural history museum across the country, had much more to offer. Despite KU's emphasis on mounting animals in recent years for the Chicago World's Fair, its mission was not all about taxidermy — the natural history museum served a much higher purpose. As part of its mission, the museum intended to create and preserve a complete record of all animals and plants on earth, both now and in the past. It therefore served as a repository of a wide variety of artifacts from nature: mounted animals nearing extinction; collection drawers of the skins of native birds, reptiles, and snakes stored in alcohol-filled glass jars; and fossilized creatures and plants that had existed millions of years ago. Collecting those objects supported the university's scholarly research and the scientific education of both students and the public in hopes of answering questions about life on our planet.

The Kansas University Museum of Natural History was established when the university first opened its doors. The original charter documents of the university specified the museum's creation, but initially referred to it as a "Cabinet of Natural History." This designation did not denote a piece of furniture; rather, the term referenced the historic "cabinets of curiosities," once-prominent private collections of rare, valuable, historically important, or sometimes quirky objects held by wealthy individuals from the Renaissance period. According to historian Wilma George, "the cabinet of curiosities was just what it said it was: odds and ends to excite wonder." Originally, these collections were treated as a trophies or as evidence of the owners' wealth and standing, and because they were privately owned, there were no boundaries on what could be found in those collections. They typically included ancient artifacts, geological objects or fossils, and specimens from nature, such as stuffed animals and preserved reptiles. Absent any overriding theme, many of those early collections contained items that were quite bizarre. Typical examples of objects in those early cabinets were "a monstrous calf with two heads" or "a horned horse." Another strange example was when Tsar Peter the Great of Russia included in his cabinet of curiosity the stuffed body of a deformed boy, dressed in a shirt and pants, whose fingers and toes had been fused together from birth and resembled "lobster claws."

The practice of displaying lurid and random oddities began to decline only when the collections were assembled for scientific purposes, like John Hunter's museum of medicine in England in the late 1700s. Hunter was a skilled and famous surgeon at the time and maintained a collection of his experiments for scientific purposes. However, items in his collection were also not for the faint of heart — such as his cutaway cross-section of a rooster head. While studying grafting, Hunter had transplanted a live human tooth in the comb of a rooster. When it was dissected several months later, the blood vessels of the rooster had accepted the tooth and were feeding it. The bizarreness of that example might be attributed to being part of a surgical experiment, but as an exhibit for public display it leaned more toward an "exciting wonder" than educational experience.

By 1895 the natural history museum at KU, like its counterpart museums across the country, showed no remnants of those curiosities and was well entrenched in serious science. In the next few years, Charles began to appreciate this modern approach to museums, and his interests pivoted from an emphasis on taxidermy to the broader subjects of the natural sciences and museum work.

CHAPTER THREE

The Museum and Its University

Wanting to make a good impression, Charles arrived at the museum early in the morning after the lecture in Topeka the previous night, despite getting home late. He was greeted by Pug Saunders when he entered Dyche's workshop in Snow Hall. Saunders was tall, ruggedly handsome, and a well-respected staff member at the museum. As the senior taxidermist for Dyche, he had been instrumental in preparing the mounted animals for the Chicago World's Fair two years earlier. Locally he was also a widely respected naturalist with a reputation as an expert hunter, fisherman, maker of nets and traps used for commercial fishing and hunting, and a builder of realistic duck decoys.

Saunders began Charles' orientation by explaining how Snow Hall was divided into four major departments. Dyche's department was called "recent vertebrates," referring to all living animal species with backbones, which included a wide range of animals: birds, mammals, snakes, amphibians, and fish. The department of vertebrate paleontology also studied the fossils of these animals, but only those from the prehistoric geological eras, for example, dinosaurs. (The word *paleontology* comes from the Greek: *paleo*, "of prehistoric times"; *onta*, "beings"; and *logy*, "study of.") Vertebrate paleontology also housed geology, the study of the earth's rocks to reconstruct its physical and chronological history, including the ages of its fossilized plants and animals. The department of entomology dealt with all insects, and the botany department with living plants. Each occupied its own distinct area that included space for its collection, classrooms, and laboratories. Such an arrangement encouraged collaboration between the faculty and students within each department. In addition, the closeness of the departments within the museum optimized cooperation between the various disciplines.

After Saunders introduced Charles to the other taxidermists and student assistants, he led him to the top floor of Snow Hall, the residence of many of the mounted wild animals from the World's Fair exhibit. With its sloping walls conforming to the exterior roof line of the building, the attic, as it was called, was hardly prime exhibition space, but it was the only space available. Once again the recent vertebrate collection had run out of room due to Dyche's prolific collecting and taxidermy activity. The crowded room contained such a wide variety of wild animals that Charles gazed with utter amazement. Suddenly he noticed one animal that seemed out of place. The tall, majestic mount of an auburn horse sporting a military saddle stood like a proud silent warrior. When Charles asked about it, Saunders quite proudly explained that the horse, named Comanche, was the last survivor of General Custer's army from the Battle of the Little Bighorn in 1876. After the horse recovered from his wounds, he became the cavalry's pampered mascot that they trotted out for special occasions. Comanche lived out his balance of days in Fort Riley, Kansas, where he died in November of 1891. Thinking the revered horse should be preserved, the army approached Dyche to do the taxidermy. According to an old note written in the 1930s by Tommy Burns, a longtime fisherman on the Kaw River, Saunders told Burns that the horse was transported to Lawrence on a railway flatcar but started decomposing when it sat on a siding for three days. When Comanche finally arrived in Lawrence, Dyche thought it useless to try to mount the horse and planned to dump the carcass on the university farm north of town. Never one to back away from a challenge, Saunders volunteered to attempt the taxidermy. According to Burns, Saunders recollected, "It was a very hard and stinking job[,] also very tedeous [*sic*] job because some of the hair had already started to slip." Many months later, when the taxidermy was finished, Comanche was spectacularly preserved. When Dyche presented his bill of $400 for his taxidermy services to Fort Riley, for reasons still unclear, they refused to pay. As a result, Comanche became part of the museum's permanent collection and accompanied the rest of the animals to the Chicago World's Fair.

Once the tour of the exhibit room in Snow Hall concluded, Saunders and Charles returned to the workshop, where Saunders pulled back the curtain for Charles to observe the inner workings of the museum. The two men began to unpack some of the many large wooden crates of the exotic specimens that Dyche had shipped back from Greenland. At the lecture the previous evening, Charles had seen the photographs of Dyche skinning the polar bears and walruses, but it did not register that most of the specimens would arrive in the museum in pieces. Considering that the museum only needed certain parts of the specimens (usually the heads, feet, skeletons, and skins), Charles thought it made good sense to leave unwanted parts in the field, especially when the animals were so large and so far from the museum workshop.

As they dug through a smelly crate of deconstructed walrus, Saunders reviewed with Charles the basic process of field dressing. He described how collectors skinned the animal, carefully separating the pelt from bone and flesh with knives and scraping tools, making sure not to damage the skin. After the cape (another term for the pelt) is removed from the carcass, the field collector preserves the skin: first, salt is liberally applied to dry the skin to prevent decomposition; next, arsenic is rubbed on the inside of the skin to prevent any insects from eating it. The crate also included assorted skulls and skeletons that had been cleaned prior to shipping. Saunders explained to Charles that the bones had been cleaned by boiling them in water and then scraped and picked by hand. While none of the primer from Saunders was new to Charles, he was impressed by the size of the specimens.

As they continued to sort through crates, Charles was awestruck by the amazing and unique animals they were working on. As a young man from the Midwest who had never seen such creatures in real life, Charles felt the experience was the chance of a lifetime. From that day forward, whenever he entered the workshop he considered himself the luckiest man in the world.

Charles noticed small handwritten tags attached by string to each of the specimens, which recorded the scientific name and date of collection. He had never observed such attention to detail back in Banning's taxidermy shop. Museum work was more methodical and precise, something that appealed to Charles. In the years to come, his aptitude for careful and detailed work led him to become a vital member of the museum staff.

Later that morning Dyche strolled through the workroom with his usual air of authority. He carefully reviewed the progress of the work and provided direction and guidance. In addition, he shared a few stories about the dangers he encountered during his hunting trip. Charles noticed that everyone in attendance gave respectful, polite attention to the professor, but always looked relieved when he left so they could return to their work. Both students and staff appreciated the professor's intelligence and energy, but also recognized a level of boasting and self-aggrandizement. Charles publicly avoided recognition and was happy to let Dyche take credit. But Saunders began to resent Dyche for reaping all the glory. For instance, in Saunders' personal account of the taxidermy of Comanche, Dyche dismissed the possibility of mounting the horse because of its condition. According to Saunders, he did all the work, but all historical reports give credit to Dyche, with no mention of Saunders.

During a break, Saunders mentioned that most animal specimens collected were never destined for taxidermy and exhibition. Saunders showed him the collection area where specimens for scientific study, rather than public exhibition, resided. Enclosed wooden cabinets contained drawers filled with the skins of birds and small mammals, each tagged and sorted by species. Those collection drawers contained thousands of smaller specimens that were just as important to the collection as the larger more exotic mammals collected by Dyche. Charles noticed that the museum collection contained numerous specimens of the same species. He would later learn that scientists were interested in species variation, which allowed them to observe and understand how animals and plants adapt and evolve over time to their changing environments.

Charles realized that his love of hunting could benefit the growing collection. To that end he began devoting weekends to collecting specimens around Lawrence. After each foray, Charles returned to the museum with his catch of birds and mammals for a tutorial from Dyche, who willingly aided in identifying the scientific name of every specimen, which were added to the museum collection. From that hands-on experience, Charles learned the entire process of specimen collecting from beginning to end, and expanded his knowledge of scientific nomenclature, a necessary skill for a competent field zoologist and curator.

Some people believe that a person's name reflects their personality. Such a notion applied to Charles when he first went to work at the museum. Initially everyone in the museum called him Charles, but with his affinity for common folks, that sounded too formal, like English royalty. Some tried to call him Charley, but that nickname never got much traction; it seemed more appropriate for someone with a gregarious and carefree personality. Certainly, Charles enjoyed a good laugh, but the attention associated with *telling* a joke made him uncomfortable. When the students called him "Mr. Bunker," that also sounded too formal for his liking. Today some people at the museum still refer to him as C. D. Bunker because of how he was listed in large group photographs. But even though that was how he signed his name on correspondence, no one at the time called him "C. D." Eventually the nickname "Bunk," bestowed by his coworkers in the workshop, seemed to stick. It was an apt name for him: simple, straightforward, and unpretentious.

In the male-dominated museum workshop, nicknames, typically a guy thing, flourished. They indicate someone's acceptance within the group while carefully reserving any real show of emotion or sentiment. That tradition of granting nicknames continued in the museum workshop for decades, with many of Bunk's students receiving their own epithet. Some were obvious, like Bunk being short for Bunker, but others were rather obscure, such as Pug Saunders'. Even today his descendants still wonder how he got that alias.

When Bunk began working at the museum, he did not fully appreciate the advantages of working for a well-respected institution of higher education. But such status was not easily earned, considering the debilitating circumstances that existed in Kansas when KU was established in 1865. At that time, Kansas had been a state for only four years, and the railroad reached less than halfway across the state. Kansans had settled only the eastern third of the state, and western Kansas was considered Indian and buffalo country, where deadly skirmishes with the Native Americans would continue for several more decades. Added to that, the Civil War raged simultaneously as the state politicians discussed establishing a state university. As for Lawrence specifically, the founding of something as grand as a university seemed terribly ill conceived when only two years earlier Quantrill's raid had killed hundreds and destroyed most of the town. With all that going on, why would Kansas and Lawrence go to the trouble to establish a new university? The answer was legislation enacted by the US government.

Prior to the Civil War, concern arose in Congress that the number of colleges and universities was not keeping pace with the growth in the newly settled states of the West. Likewise, the residents of the West desired the same cultural and educational opportunities that they remembered from back east. To remedy that, in 1862 Congress passed the Morrill Act, also known as the Land Grant College Act. The law granted federal land to every state, hence the name of the act, for the purpose of establishing colleges and universities to specialize in "agriculture and mechanical arts." The actual acreage gifted to a state was based on a formula related to the number of their senators and representatives in Congress. After they chose their acreage, states could either use it for the university campus or sell it and use the proceeds to purchase a more suitable property. In exchange for the land, Congress stipulated that those universities be required to offer degrees in agriculture and engineering, hoping those institutions would ultimately contribute to a state's economic development.

Although another point in time might have allowed for the university to be established in a more organic manner, once that law was enacted, every state in the Union, including Kansas, leapt into action. Immediately Kansans became embroiled in political infighting over where to locate their land-grant university. Communities wanted the land-grant school, thinking it would make their town more viable for the long term, similar to being designated as county seat or located on a railroad line or by a river. When the political dust finally settled in 1863, the land-grant college was awarded to Manhattan and named the Kansas State Agricultural College. However, as a political compromise, the normal college (a historical term for a teachers' college) would go to Emporia, and Lawrence was designated to be the home of the State University of Kansas.

Before the final decision was made by the legislature, the city of Lawrence lobbied to be the location of the land-grant school by purchasing a forty-acre parcel of land for the university where Fraser Hall and Snow Hall would eventually be built. The plot was southwest of town and situated on the top of a hill known locally as Mount Oread, a name given by the first settlers to honor a seminary from back home in Massachusetts. However, in 1865, when it came time to begin constructing the first university building, local concern arose that maybe the forty acres was just too far away from town. Three blocks to the north of that parcel was an eight-acre tract owned by the local Presbyterian church, slightly closer to town, but still on top of Mount Oread. The church owned the property with the intention of building a denominational university and had even completed a fifty-by-fifty-foot foundation for the structure. However, a change of heart by the church elders caused them to sell the acreage to the city fathers, who donated it to the state for the new university. Immediately a two-story structure was erected on the site that overlooked the town of Lawrence. When the building was completed in time for the fall term and a three-person faculty was hired, "University Building," later renamed North College, opened its doors to students for the first time in 1866.

One of the original three faculty members, Francis Huntington Snow stood only five feet, five inches tall and appeared younger than his years. Consequently, he was often confused for a student, which may be why he sported unusually long mutton-chop sideburns for much of his career. First impression aside, it soon became obvious that he was a notable scientist, a tireless worker, and an inspiration to students. Frank Snow always referred to KU as a "first-class university." From the very beginning he wrote to his fiancée that the faculty held firmly to the belief that the "standards [at the university] were as high as at Harvard." Despite Snow's confidence in academic standards, however, the first students at KU arrived unprepared for the rigors of college. According to Clifford Griffin's history of the university, "in all the ways that mattered the so-called University of 1866 was merely a preparatory school for a nonexistent college." Snow accused state politicians at the time of getting ahead of themselves by establishing universities when there were "no high schools in the state outside of the cities." As a result, the new university faculty for the first two years taught only a high school curriculum to prepare the students for college-level coursework. It would be more than two decades before enough high schools existed across the state to prepare students for college and thus eliminate the need for preparatory classes at KU. Also, because about 90 percent of the first KU students came from Lawrence and the surrounding Douglas County, critics began referring to the university as the "Lawrence High School."

Like Bunk, Snow was a Kansas transplant. He was born in 1840 in Fitchburg, Massachusetts, a progressive eastern community with a reputation for opposing slavery, much like Lawrence. Snow did well enough in school to earn admittance to Williams College, in Williamstown, Massachusetts, where he studied Greek and science. Prompted by his interest in the natural sciences, he joined a science organization called the Lyceum of Natural History. Snow collected and discussed natural science specimens with the other members, and in 1861 they voted him president. In 1863, while he was still in school, the Union army drafted him, but although he supported the North's opposition to slavery he did not believe in war and paid $300 to avoid serving. When Snow passed his examinations in Greek and Hebrew at Andover Theological Seminary on August 1, 1864, he set out for Washington, DC, to become a delegate of the United States Christian Commission. The organization supplied the armed forces with physical comforts not furnished by the government, to strengthen their morale, and ministered to their spiritual welfare.

Shortly after the war Dr. Charles Robinson, an old family friend from his days in Fitchburg, contacted Snow. Robinson had moved from Fitchburg to Lawrence with the original antislavery settlers, was later elected governor of Kansas, and now served on the fledgling university's board of regents. Dr.

Robinson wanted Snow to come to Kansas to teach language, mathematics, and science at the new university to be established in Lawrence. Kansas was thousands of miles away from Massachusetts and on what Snow likely feared was the edge of civilization. However, Snow greatly admired Dr. Robinson and was eventually persuaded to accept the position, relishing the opportunity to build a university from the ground up.

In the mid-1800s Harvard University revolutionized the teaching of the natural sciences with a new method called "the direct study of nature." This practical, hands-on system involved students spending time in both the field and the museum workshop collecting, preparing, and storing the specimens for the museum. Two decades later, when the University of Kansas first opened its doors, Frank Snow made sure that KU adopted the same practice. For the first ten years Snow found it difficult to balance classroom lecturing and fieldwork, especially with the demands of teaching preparatory classes. As a result, fieldwork was limited to the area near Lawrence. By 1876 Snow recommitted to the direct study of nature by expanding scientific expeditions to outstate Kansas, Colorado, New Mexico, Texas, and Arizona. Such trips usually lasted at least a month or more during the summer and included both students and faculty from various disciplines. They focused on collecting a wide variety of items, everything from fossils, rocks, plants, birds, mammals, and reptiles to insects, which were Snow's favorite. During the school year, supervision of the students was relegated to the museum staff, like Saunders. Years later, after Bunk joined the museum, he would become involved with training the students in fieldwork, something that would later distinguish his career.

Frank Snow always regretted the necessity of teaching students who were ill prepared for the college curriculum, but despite that concern, Dyche appreciated those preparatory classes because he could not have enrolled at KU in 1877 without them. After Dyche enrolled in precollege classes, he flew through his coursework, excelling to the point that in 1882 he began teaching anatomy and zoology classes even though he was still in his junior year in college. Not only did Dyche stand out in the classroom, he displayed an extraordinary aptitude for fieldwork. His first foray on those summer expeditions occurred in 1878 while he was still an undergraduate. Using his childhood experience as a frequent hunter and camper, field collecting came naturally to him.

By 1890 many of the faculty had earned their advanced degrees at prestigious universities back east or in Europe and arrived at KU sporting their

Phi Beta Kappa keys. As a result, they pushed to establish the first chapter of that esteemed organization west of the Mississippi River. To offer perspective, this designation would not occur at the University of Michigan for another seventeen years. To raise the bar for the sciences, KU also established a chapter of Sigma Xi, a prestigious scientific honor society, only the fourth in the United States, the other three being universities on the East Coast.

The year 1890 marked the beginning of a watershed period for the University of Kansas. That was the year KU needed to search for a new chancellor. The regents wanted Frank Snow as the next chancellor, but as a condition of his acceptance, he wanted to enhance the status of the natural sciences and insisted that they hire a world-class scientist, Dr. Samuel Wendell Williston. Williston was an imposing figure with broad shoulders and stellar academic credentials. After earning both a medical degree and a doctorate from Yale University, he joined the Yale faculty and taught anatomy. In addition, he was a well-respected paleontologist and entomologist, specializing in Diptera, the taxonomic category that includes flies, gnats, and mosquitoes. The day after Williston accepted the position at KU, Snow agreed to become the chancellor.

Hiring Wendell Williston away from such a prestigious university as Yale to join a faculty so far away from the comforts of the civilized East Coast may have seemed to some an unusual achievement. But Williston had a fondness for and a long relationship with the state of Kansas. In 1857 he and his family settled in Manhattan, Kansas, located seventy-five miles west of Lawrence and home to the Kansas State Agricultural College. Coincidentally, the Williston family arrived as part of a colony of antislavery immigrants, the same organization that brought early settlers to Lawrence. As a young boy Wendell excelled in school and eventually became acquainted and worked with Benjamin Franklin Mudge, who was a professor of natural history at the college in Manhattan. Through that connection Williston was hired to collect fossils in the rich fossil beds of western Kansas for Othniel Charles Marsh, the famed director of Yale's acclaimed Peabody Museum. Although O. C. Marsh was raised in poverty, a wealthy uncle established the museum and arranged for Marsh to direct it. Over time Marsh successfully amassed a large collection of prehistoric fossils. Williston was encouraged to attend Yale after working for Marsh for several years, but by 1890, when KU approached Williston about coming to Kansas, Williston and Marsh had had a serious falling out over Marsh's careless handling of the payments due Williston. Williston also resented Marsh's habit of not sharing scientific credit with those who worked for him. Additionally, he remembered the fossil-rich chalk bed of western Kansas that preserved an abundance of valuable marine fossils: mosasaurs, ichthyosaurs, pterosaurs, fish, crinoids, and many others. The prospect of KU's dedication to building up a fossil collection, the proximity of those fossil fields, and his desire to distance himself from Marsh all contributed to his decision to come to KU.

When Bunk joined the staff of the natural history museum, paleontology did not conjure up the same visceral feelings that he felt for living birds and mammals. Dinosaurs (the name for which, coined by the English scientist Richard Owen in 1841, came from the Greek words *deinos*, meaning "terrible," and *sauros*, meaning "lizard") looked monstrous and bizarre when compared with the species now living in nature. However, most of the museum naturalists in those early days were Renaissance men: they had a focused interest but dabbled or moved between different scientific disciplines as they followed the sometimes irregular path of scientific questions. Such an intellectual environment motivated Bunk to become acquainted with the world of fossils and fossil collectors and their activities. He learned that fossil remains enable scientists to study and reconstruct the history of life on Earth, how animals and plants have adapted and evolved over time.

As a serious scientific inquiry, paleontology began after Darwin published his *On the Origin of Species* in 1859. From that point forward, every discovery of a new fossil contributed to an understanding of evolution. However, even the most astute scientists don't always agree. For example, Williston learned about evolution from classes with Mudge, who remained opposed to Darwin's theories until his death. Williston, on the other hand, through his study of the fossil record, became convinced that evolution explained the diversity of life on Earth.

The earliest discoveries of prehistoric fossils in the United States dated back to the 1700s, when fossil hunting developed "as a gentlemanly pastime of the East Coast." For example, one of the nation's founding fathers, Thomas Jefferson, published a treatise on vertebrate paleontology in 1797. Later in the 1870s, fossil collecting changed from a dignified activity to a competitive sport spurred by two American paleontologists: Edward Drinker Cope and the aforementioned O. C. Marsh. They engaged in an intense competition to see who could build the finest and largest collection of fossils in the country — Cope for the Academy of Natural Sciences in Philadelphia, and Marsh for the Peabody Museum at Yale University in New Haven, Connecticut.

Both men were fixated on being the first to find—and, more importantly, announce—various new species, as well as the oldest fossils of a particular species. In his race for fame, Cope was more willing to speculate about his discoveries, which could lead to mistakes. The most famous example of such an error occurred in 1869 when Cope assembled a skeleton of the late Cretaceous (the geologic time period from 145 to 66 million years ago) plesiosaur *Elasmosaurus platyurus* discovered in western Kansas. Quickly after he finished, he published an article without any peer review, only to be mortified when it came to light that he had placed the skull on the wrong end.

Although Cope and Marsh did their own fossil collecting early in their careers, they both eventually realized that hiring independent collectors could increase the rate of growth of their fossil collections. Marsh's team included Mudge and Williston, while Cope employed Charles Hazelius Sternberg (who lived in Lawrence, but spent most of the year collecting fossils in western Kansas). Naturally that rivalry between Cope and Marsh carried over to the members of their teams. One factor in the race to grow the largest collections involved secretly guarding the locations of their fossil digs. One example of the cloak-and-dagger relationship between the rival teams occurred in 1877. Sternberg recounted the time when he and Williston met unexpectedly on the same westbound train: "I was surprised to see Mr. S. W. Williston get aboard with his outfit [for fossil collecting] . . . when he entered my car, he was greatly astonished, thinking that I was on his trail. He tried to find out my destination, but failed."

Today the legacy of Samuel Wendell Williston at the University of Kansas remains significant. Among his major contributions is the wealth of fossil specimens he collected for the KU museum from the Cretaceous Niobrara Chalk — a geologic formation found in western Kansas and Nebraska — during his twelve-year tenure. But he also left something else, lesser known but likely just as important: his commitment to training young scientists. After Bunk arrived in 1895, he observed how Williston had assembled a group of talented students who were extremely loyal to their "beloved teacher." Not only did his students recognize his vast intellect, they felt his greatest asset was his ability to personally connect with them. Williston truly cared about his students and looked out for them. For instance, one student could not afford to live on his own at KU, so Williston invited him to live with his family for an extended period. In addition, the students considered Williston a regular guy with human foibles. According to his biography by Elizabeth Shor, *Fossils and Flies: The Life of a Compleat Scientist, Samuel*

Wendell Williston (1851-1918), "He kept a cluttered office and was often distracted — he once went into a classroom to quiet the pandemonium only to discover it was his own class that he had forgotten — but these qualities gave his students a sense of his humanity and made him endearing."

Williston also related to the common man. According to one commentary, Williston "preached a democracy of science where neither wealth nor status mattered so much as intellect and hard work. His students all wanted to be some version of him, and in so doing, their success became a testimony to the power of that message." During their time together at the museum, Williston's authentic personality impressed and influenced Bunk. By comparison, Dyche often came across as aloof and self-absorbed. He also demonstrated that his students were not his priority by frequently paying substitutes to teach his classes so he could engage in big game hunting, give public lectures, or lobby legislators. As a result, the students and staff respected Dyche for his skills, knowledge, and accomplishments, but they never revered him like they did Williston.

One notable student who excelled under Williston's encouragement was Clarence E. McClung. Only a few months older than Bunk, McClung entered KU to study pharmacy in 1890. By 1892 he had earned a degree in pharmacy, but elected to change fields from chemistry to zoology with Williston's encouragement. McClung, who was exceptionally bright and gifted, earned a degree in 1896 and taught zoology as an assistant professor for several years. By the end of the century, through his research on chromosomes in insects, McClung was one of the first to offer evidence that chromosomes carry certain hereditary traits that determine sex. His work eventually attracted worldwide attention.

Another talented student of Williston's, who attended KU during Bunk's tenure, was Barnum Brown, born in 1873 and named after the famed circus promoter P. T. Barnum. His name prophetically described Brown, who later admitted, "There must be something in a name, for I have always been in the show business of running a fossil menagerie." He grew up in Carbondale, Kansas, where his father operated a freight wagon company for the government before the railroads spread across the Midwest. Brown's father even took him on a long wagon trip to Yellowstone National Park in 1889 to "show him the world." However, when trains began to replace freight wagons, his father turned to farming and selling coal that he mined from his farm to the railroads. In 1893 Brown enrolled at the University of Kansas to study engineering, but soon came under the wing of Professor Williston. In 1895 Williston organized an expedition to the plains of Wyoming to search for the fossilized bones of a *Triceratops*, and needed someone to drive the supply wagon to the fossil field in advance of the collection party that would travel by train. Because of Brown's experience with freight wagons and knowledge of the trail to Wyoming, Williston asked his young student to join the collecting party.

Brown turned out to demonstrate a natural talent and fascination for fossil hunting, which suited him better than the classroom. In 1896 he left KU before earning a degree to work for the American Museum of Natural History (AMNH) in New York as a fossil collector. He got the job with Williston's recommendation, but quickly gained a permanent position a few years later. Despite his scholastic shortcomings in the classroom, Barnum Brown became known as one of the greatest dinosaur collectors of his time. Later he would be credited with discovering the first *Tyrannosaurus rex* in Hell Creek Canyon in Montana in 1902. The *T. rex* was one of the largest meat-eating dinosaurs, standing up to thirteen feet tall and forty feet long. According to Lowell Dingus and Mark A. Norell in their book *Barnum Brown: The Man Who Discovered* Tyrannosaurus Rex, in addition to his reputation as a noted paleontologist, "Barnum Brown lived fast and worked hard. He was a wilderness gourmand, a natty dresser, a drinker, gambler, and smoker."

Another notable student who came under Williston's wing was Elmer Samuel Riggs. Despite never having taken a course from Williston, Riggs joined the 1895 *Triceratops* trip to Wyoming because he had distinguished himself when he worked on a fossil trip to South Dakota the previous year. Unlike Brown, Riggs left KU with a degree and joined Brown on the 1896 expedition from the AMNH. Riggs later did graduate studies at Princeton University before getting a job at the Field Museum in Chicago, where he embarked on a forty-four-year career.

In 1899 Handel T. Martin arrived at KU, not as a student but as a field collector and fossil preparer. Small and wiry, Martin possessed inexhaustible energy. Like Bunk, Martin lacked a formal education past grammar school. He had been born in England, where he learned about fossils from a young student home on vacation from Oxford University. As a young man he migrated to the far western part of Kansas to work as a ranch hand and hunt fossils in his spare time. During the 1880s Williston hired Martin as an independent contractor on several of those western Kansas fossil digs for Marsh at Yale.

In 1895, the year Bunk was hired by Dyche, a farmer in Logan County, Kansas, found some bison bones protruding from a bank along Twelve Mile Creek of the Smoky Hill River. Thinking these bones might have some scientific value, he contacted Williston at the University of Kansas, who dispatched Martin and assistant T. R. Overton to examine the bones. What they discovered was a prehistoric bison skeleton, which later became known as the Twelve Mile Creek Bison. Although the fossilized skeleton was considered one of the university's most notable finds, a shroud of mystery continues to surround it today. As the story goes, when Martin excavated the fossilized remains, he discovered a fluted stone point impressed against the shoulder blade of the bison, indicating that humans had hunted the animal. Carbon dating later determined that the bison skeleton was 10,300 years old, proving by association that the arrowhead was the same age and evidence of the earliest man found in North America. Martin's arrowhead would have been a phenomenal discovery had it not been for an unexpected incident. A few years later Williston supposedly attended an evening gathering at the chancellor's residence. Such get-togethers, where prominent townspeople were invited to mingle with the university faculty and the faculty had the opportunity to share their interesting discoveries, occurred on a regular basis. On this occasion, Williston had prepared to discuss the Twelve Mile Creek Bison and show off the arrowhead. According to Martin, during his presentation Williston passed the arrowhead around the room. At the end of the evening, when he tried to retrieve it, it had mysteriously disappeared. To this day the arrow point has never surfaced to offer proof positive of its existence. Without the actual arrowhead, the claim of evidence of the oldest example of man in North America could never be proved.

After Martin was hired at the university, he and Bunk became fast friends and allies. Martin enjoyed hunting and frequently joined Bunk in collecting specimens of birds and mammals around the outskirts of Lawrence. Through this relationship, Bunk became acquainted with fossil collecting and Martin learned about collecting birds and mammals. Because both men were self-taught scientists with limited formal education, they stood out as anomalies within the university setting. Consequently, they soon learned to watch each other's back.

CHAPTER FOUR

Taxidermy as a Solution for the Dilemma of Extinction

When the University of Kansas Museum of Natural History first opened its doors, exhibition was central to its mission. Exhibition serves three functions. First (not necessarily in the order of importance) is public relations — to report to the taxpayers of the state (and the legislature who controls the purse strings) what is going on in the university, and to demonstrate that the money from the state is being well spent, which in turn justifies future financial support. Second, to excite the public about the wonders of nature. Such a message engages children with natural science so that work at the museum will continue in future generations. Third, and probably the most imperative, the museum must inform and educate the public of worries that the scientific community holds about the health and future of the life of the planet.

Bunk's introduction to the mounted animals in the museum attic exposed him to the latter: the role of the museum to warn the public about endangered species. In the 1870s most scientists across the country shared the concern that certain endangered species would become extinct before they could become part of the physical record of the Earth that resided in natural history museums. Throughout the history of the Earth, species extinction occurred naturally as part of evolution. However, in the nineteenth century the number of species becoming extinct was increasing at a rate directly related to the growing human population spreading west across the United States. That worrisome phenomenon even affected some species whose population numbers were previously considered invincible; for example, the bison, commonly called the buffalo. At the time of the Civil War, anywhere from thirty to over sixty million bison, depending on the source of the information, roamed on the plains of North America. When the railroads moved west onto the vast prairie, the buffalo were heavily hunted as food for the railroad workers. In addition, their skins and other parts were shipped back east for clothing and household supplies. Moreover, a policy of the United States government also directly contributed to the dramatic decline of the buffalo. In order to encourage settlement of the land west of the Mississippi River, the government tried to move Native American tribes off the land they occupied by utilizing tribal treaties. When those treaties broke down or the Native Americans refused to negotiate, the conflict over control of the prairie often resulted in bloody fighting. When the government doubled down and assigned military troops to protect the settlers, the battles only escalated. Enter the buffalo. The military leaders of the government troops reasoned that the buffalo were integral to the survival of the Native Americans, so if bison were eliminated, the Native Americans would stop fighting and be more willing to move to reservations. Evidence of that position was supported by a quote attributed to General Philip Sheridan: "Shoot a buffalo, starve an Indian." Because of that thinking, soldiers, sportsmen, and professional hunters were encouraged to indiscriminately massacre large herds of buffalo to the point that by the late 1880s there were no buffalo roaming the western prairie. Bison would have most likely become extinct had it not been for those held in captivity.

Another example of endangered species was related to the improved transportation in the late 1800s. Prior to the railroads connecting the Atlantic and Pacific Oceans, the remote mountain ranges of the West were home to a multitude of large mammals: moose, elk, cougars, bear, mountain goats, and bighorn sheep, to name a few. Their population numbers were primarily protected from sportsmen by their remoteness. However, as people settled farther west, the populations of many of those wild animals began to decrease when they were killed for food or sport or their habitats were disturbed or destroyed by encroaching civilization. When the railroads eventually crossed the country, trains delivered hunters from the East to the environmental doorsteps of those endangered animals of the Rocky Mountains. As those animals continued to show up as trophies above fireplace mantels, their populations continued to decline, further exacerbating the problem for natural history museums.

In the 1880s Snow and Dyche pondered the rapidly increasing problem of species disappearing. Resigned to the impossibility of stopping the country's population growth from moving west, like other museums across the country, KU's museum embarked on a Faustian plan. Dyche would hunt and kill those endangered species before they became extinct to make sure to at least preserve a physical record of the species. His rush to collect endangered species began in earnest in the 1880s, when he began his expeditions to collect big game mammals that occupied the remote areas of North America. Thinking about how many species the museum needed to exhibit, Dyche began rethinking KU's approach to taxidermy.

Previously the museum had utilized taxidermy sparingly because most scientific research could be done from the skins and skeletons. On those occasions when they wanted a specimen mounted, they engaged a private taxidermist in Kansas City named Frank Dixon, who Dyche felt had limited skills. That opinion resulted in the decision to bring the taxidermy activity in-house, and in 1883 Dyche made his first foray into taxidermy with several wild turkeys and a golden eagle that he had shot the previous year.

About that same time, hundreds of miles away in Mendota, young Charles began observing Mr. Banning mounting birds in his taxidermy shop. At the time, Bunk considered Banning an admirable taxidermist, but after observing the taxidermy at the KU museum, he understood the difference between Banning preparing a trophy and a museum taxidermist mounting a specimen. According to historians of taxidermy, commercial taxidermists seek to preserve the memory of a successful hunt. The mount need not look exactly like the particular animal killed, but rather look generally like the animal. Also, for a taxidermist like Banning, the object was a commercial transaction, which encouraged him to work quickly. By comparison, museum taxidermists took all the time necessary to accurately depict the specific animal, taking into consideration the latest phylogenetic or evolutionary information of the specimen.

Fortunately, in the early 1880s both Dyche and Banning were working on birds whose feathers covered or hid the anatomy. Mammals, on the other hand, offer more of a challenge because in many cases all of their musculature is visible. For Banning, specializing in birds allowed him to make a living. But Dyche, who knew that many of his endangered specimens would be big game mammals, struggled for several years refining his methods to make them look real and alive. During that period, he attempted several larger mammals, a mule deer and a bear, but wasn't completely satisfied with the results. Despite that, he continued to hunt and stockpiled many of the larger animals for when he was more confident in his skills as a taxidermist. As Dyche became acquainted with recent examples of high-quality taxidermy across the country, he became determined to find someone who had perfected it as a fine art.

The gradual rise in the quality of taxidermy in the United States began during the Civil War. In the 1860s museums and university collections across the country increasingly sought display specimens to exhibit to the public. To keep up with high demand, natural history museums often turned to outside sources, such as Professor Henry Augustus Ward. In 1862 he opened Ward's Natural Science Establishment in Rochester, New York, to supply everything from fossils and skeletons to mounted animals that were prepared by a staff of taxidermists. In its early years much of the taxidermy at Ward's was so inferior that it earned a reputation as a "taxidermy factory." Carl Akeley, a highly talented taxidermist who had been an early graduate of Ward's but later enjoyed a stellar career at the Chicago Field Museum and the American Museum of Natural History in New York, wrote that in the early days of Ward's they took an animal's skin from a hunter or collector and stuffed it or "upholstered" it (a fitting rebuke considering that some of the very earliest examples of taxidermy were done by makers of overstuffed furniture simply because they knew how to run a large needle through a tough piece of fabric). Furthermore, he condemned their efforts by saying that they filled "a raw skin with the greasy bones of the legs and skull and stuffed the body out with straw, excelsior, old rags and the like." (Incidentally, the nationally renowned Field Museum was established when the retail tycoon Marshall Field purchased the entire Ward's Natural Science Establishment exhibit at the Chicago World's Fair in 1893 for $100,000.)

In 1880 the quality of taxidermy began to show signs of improvement. That year the Society of American Taxidermists held its first taxidermy competition, "to elevate taxidermy from amateurish craft to a professional-grade fine art by proving it could be both scientifically sound and artistically evocative." Despite their lofty stated goal that year, many of the mounts displayed were still crude. However, one caught the eye and imagination of everyone in attendance. It had been created by William Temple Hornaday, who had also worked at Ward's during the early days and was then the chief taxidermist at the National Museum (the natural history museum that is part of the Smithsonian Institution) in Washington, DC. In his famous *A Fight in the Tree-Tops*, Hornaday skillfully re-created a fight between two male orangutans swinging from jungle vines. To add to the drama of the scene, one had bitten off the finger of the other, whose face "writhed in agony as blood oozed from his mutilated hand." Hornaday recognized the importance of showing the specimen in a tableau that represented how the animal lived in its natural habitat.

From that point forward, Hornaday became known as one of the preeminent taxidermists in the country. His work developed into a highly evolved art form whose chief objective was to freeze motion to allow the observer to see an animal up close and frozen in time, something that otherwise would be impossible to see in nature. It was that kind of taxidermy that led Roy Chapman Andrews, paleontologist and well-known explorer at the time, to say, "If you want to see a live elephant, you can go to a circus or a zoo. But if you want to see the way an elephant lives, you go to a good museum of natural history."

While Hornaday gained notoriety for his skill in mounting animals, Dyche continued searching for someone to teach him the finer points of taxidermy. The problem was that only a handful of truly talented and artistic practitioners existed nationally, and they were all located back east, far from Kansas. Like many of the famous paleontologists of the era, the truly exceptional taxidermists were very competitive and closely guarded their secret techniques. So Dyche had no one to help him carry out his vision of creating a collection of endangered mammals of North America until a series of events that occurred in 1886.

That year marked a watershed period for the University of Kansas Museum of Natural History, when the Cabinet of Natural History blossomed into a bona fide museum. With Dyche's success collecting so many larger mammals, the space allocated for the natural sciences in Fraser Hall became increasingly overcrowded. To rectify the situation, Frank Snow approached the state legislature to fund a new building for the natural sciences. When they agreed to pay $50,000 for a new building, construction began immediately. To honor the beloved professor, Snow Hall opened in time for the fall semester of 1886.

With this move, the university could shed the name "cabinet" for a more prestigious designation: Museum of Natural History. (The word *museum* comes from *muse*, meaning to inspire or provoke thought.) The new museum was dedicated on November 16, 1886, with the noted paleontologist Edward Drinker Cope giving the principal address.

At the same time Snow Hall was being dedicated in Lawrence, hundreds of miles away in a remote region of Montana events began to unfold that would eventually lead to Dyche learning from one of the country's most successful taxidermists. Hornaday shared the same concerns as both Snow and Dyche. He saw the bison vanishing from the western prairies and wanted to collect specimens before the species became extinct. He called his efforts "war for wildlife." In 1886 he organized an expedition for the National Museum that he sarcastically called the "Last Buffalo Hunt." The destination for the hunting party was Montana, in an area that would eventually become part of Yellowstone National Park.

One of the members of that hunting party was a twenty-four-year-old University of Kansas student named William Harvey Brown. Snow had arranged with Hornaday for Brown to work that summer to gain museum experience. When Snow learned of the intended Montana buffalo hunt in the fall, he offered the services of his student and a sum of money in exchange for a bison specimen for KU. The Smithsonian expedition left Washington, DC, in late September; when they finally broke camp on December 15 the hunt had produced twenty-two fresh buffalo skins, forty-four skulls, eleven skeletons, and various skins and bones collected along the way.

In March of the following year KU not only received the promised bison, but a second specimen. With their arrival, Dyche saw in Brown's relationship with Hornaday a way that he might get training with a truly gifted taxidermist. Dyche asked the young student to reach out to Hornaday to see if he would agree to show Dyche how to mount the buffalo. Hornaday initially refused, but eventually agreed in April of 1887. Because Hornaday was beginning to work on mounting the last of the six buffalos he was going to exhibit, he sent instructions for Dyche to come to Washington immediately.

For two weeks, Hornaday provided Dyche step-by-step training on how to mount a buffalo. Dyche absorbed all the techniques that Hornaday had developed — everything from building the structural support to special techniques for preparing eyes, nose, and ears. The first step involved constructing a frame to replicate the skeleton using both bones from the animal and wood when bones were not available. To that frame, Hornaday showed Dyche how to attach the animal's skull. Next, slender pieces of wooden lath were fastened to the frame, and then excelsior — wood shavings twisted in the shape of a rope — was wrapped around the lath frame. Modeling clay would be applied over that, which Hornaday shaped and molded to replicate the musculature of the animal. He impressed on Dyche just how important an extensive knowledge of anatomy was for the final mount to look lifelike. After the manikin was completed, the tanned hide was tightly stretched over it, resulting in an exceptionally realistic-looking buffalo. According to Hornaday, a skilled taxidermist was "a combination of modeler and anatomist, naturalist, carpenter, blacksmith and painter. He must have the eye of an artist, the back of a hod-carrier, the touch of a wood-chopper one day, and an engraver the next."

After Dyche completed his two weeks of training, he spent most of the summer touring back east. In addition to visiting family, he stopped at several museums to check out their taxidermy and concluded that the work of most museums fell short of Hornaday's and was certainly not good enough for KU. With his new knowledge and expertise, Dyche immediately began working on mounting KU's new buffalo. He felt confident that he could successfully create the largest menagerie of endangered mammals of North America, an extraordinary collection that he would eventually take to the Chicago World's Fair in 1893.

That was the background for the taxidermy training Bunk received once he came to KU. From the fall of 1895 until the end of the century, Bunk worked diligently to refine his skills as a collector, preparator, and taxidermist. However, as the new century began, the heyday of taxidermy was past, in particular for the big game mammals. Once the endangered species were made part of the physical record of the museum collection, natural scientists turned their attention to filling the collection drawers. The days of the expensive grand adventures to all corners of the continent were past. Bunk's exposure to the history of KU, the museum, and the art of taxidermy allowed him to understand and appreciate what a grand opportunity he had been given and that it was up to him to take advantage of it. He loved working at the natural history museum at the University of Kansas and truly respected and admired the faculty, administration, staff, and students. However, as the century came to a close, an opportunity presented itself that made Bunk think maybe the grass might be greener on the other side of the fence.

CHAPTER FIVE

Oklahoma or Bust

On June 10, 1901, Bunk boarded the early morning train to Norman, Oklahoma. Two days earlier he had received a letter from Dr. Albert Heald Van Vleet, a well-respected faculty member at the University of Oklahoma and the state geologist for the Oklahoma Territory (Oklahoma would not become a state until 1907). The letter informed Bunk that he had been elected to serve as the collector and taxidermist for the university. The Oklahoma Territorial legislature had recently created the position when they established a geological survey and natural history museum. They charged the new organization to survey, collect, and/or identify all plants, animals, and geologic formations in Oklahoma. They wanted Bunk to create a new collection of zoological specimens for their university because during his time at KU he had quietly earned an excellent reputation as a field zoologist and collection preparer. Although he was grateful for his time at KU and indebted to Dyche for introducing him to the natural history museum world, he accepted the offer. He believed it was an opportunity to build something from the ground up, just like when Snow had accepted the faculty position at KU. He hoped a new institution would give him a chance to elevate his professional status, which had always been a challenge while working under the shadow of the dynamic Dyche.

Bunk found an isolated seat in the back of the passenger car and settled in for a long ride that would not reach Norman until after midnight. The noise and motion of the rumbling train reminded him of his trip by rail to Kansas with his parents ten years earlier. Just as then, he knew little about either his destination or the people who hired him. However, this time he felt honored to have been chosen for the new position. This trip was also different because he was traveling alone; his parents were settled in Lawrence. However, this time he would share this new adventure with his wife, Clara, who would follow him to Norman once he found a place for them to live.

Bunk first became acquainted with Clara Parnell at the Plymouth Congregational Church in Lawrence, where both of their families attended with some regularity. For Bunk's family, religion had always been important. His grandfather, Slocum Bunker, grew up in a Quaker family and was a loyal member of the Society of Friends for many years. Even though Slocum held leadership positions in the Quaker Church, when he became active in local politics and was elected mayor, the church disowned him. His life had become too secular, and that was contrary to the tenets of the Society of Friends. Eventually Slocum joined the Presbyterian Church, which contributed to David Bunker leaning toward the more liberal protestant churches during his married life. Regardless of where David stood on the religious spectrum, his interest in religion was acknowledged in the 1892 Haskell Institute superintendent's report to Congress. The report stated, "There is a Young Men's Christian Association properly officered and this, under the direction of its officers, has frequent prayer meetings. Mr. David Bunker, the wagon-maker, has, with commendable kindness, conducted a volunteer Bible class on Wednesday evenings during the year."

Clara Parnell was pretty and petite, with dark curly hair and a pleasant smile, and like Bunk, she was shy and quiet. Her father, A. J. Parnell, a well-respected community leader, had been elected to serve on the Douglas County Commission. (Incidentally, he was named after a notable family relative, President Andrew Jackson.) Clara's family successfully farmed a considerable amount of land around Lawrence, so many considered the family well-off. Regardless of the family wealth, Clara had grown up living a comfortable but simple and unpretentious life. When Bunk and Clara became better acquainted, she arranged with her family for Bunk to hunt on the Parnell farm ground, which always gave him the chance to stop at the house to see Clara. Over time their friendship gradually blossomed into a nearly two-year courtship, and they were married in the Congregational Church on January 19, 1898.

Following their marriage the newlyweds moved into a small two-story house in the 1700 block of New Hampshire Street, just a block from Bunk's parents. Living on their own and with Bunk still not making much money at the university, the young couple barely made ends meet. Their lives consisted of Bunk working long hours at the museum throughout the week and then driving out to the family farm on the weekends to cut firewood to sell and use. He often took his gun just in case he had an opportunity to collect bird and mammal specimens for the museum. Like most young brides at the time, Clara managed the home, including the cooking, cleaning,

Charles D. Bunker circa 1898 (University of Kansas Natural History Museum records, University Archives, RG 41/0 Photographs, Kenneth Spencer Research Library, University of Kansas Libraries)

Clara Parnell Bunker circa 1898 (family photo)

laundry, and managing the household finances. During the summer she tended a sizable garden with vegetables and fruit for canning. Those first years of marriage caused Bunk to ponder what he could do to improve their financial situation. Occasionally he thought that maybe his position at the museum might be holding him back.

When the letter from Van Vleet arrived on Saturday, June 8, Bunk was elated. He and Clara discussed the merits and shortcomings of the job offer. Clara's first thought was that she didn't want to move so far away from family. Although Bunk agreed, he was convinced that it was a great chance to be his own man. Once they arrived at their decision, Bunk wired Van Vleet accepting the position. Because the job offer specified that they wanted Bunk to join a collecting trip that was scheduled to depart on June 11, he hurriedly began to prepare to leave Lawrence on Monday, June 10.

The early morning train left Lawrence, heading west to Topeka. There Bunk changed to another train heading south. Bunk knew that it was going to be a long day on the train, but his excitement about his new job helped set him at ease. As the train steamed southwest, it began to cut through the scenic Flint Hills of Kansas. The undulating treeless prairie resembled oversized ocean waves. He found it mesmerizing. As he stared at the beauty of the landscape, he became lost in his thoughts. Bunk reflected on why he was leaving KU, when five years earlier he had considered it his dream job. He had fond memories of his time at the museum and appreciated all that he had learned there. But recently he had begun to worry that the sands at the university were beginning to shift. Maybe the museum might be changing, and not for the better.

Chancellor Francis Huntington Snow led the University of Kansas to new heights during the 1890s. He assembled a distinguished and dynamic faculty, in every area of the university, not just in the sciences. Those scholars arrived at KU with lofty aspirations, intent on supporting high academic standards. The museum of natural history experienced extraordinary growth in the number of fossil specimens added to the paleontology collection due to the efforts of Professor Williston. Also, the KU zoological exhibit of North American mammals that Professor Dyche presented at the 1893 Chicago World's Fair gave the museum a national reputation for the very first time. Those two accomplishments had resulted in discussions about the need for a new and larger museum of natural history. But despite all that success, Bunk and others shared concern about the future of their museum when it was officially announced that Snow had stepped down as chancellor. His departure was not a surprise, since they had seen his health decline from years of constant warring with the state legislature. One area of contention related to KU teaching Darwin's theory of evolution, which infuriated the religious conservatives in the state capitol. Even though Snow had personally reconciled religion and evolution and was less dogmatic about evolution than other faculty, as the head of the university he became the lightning rod for their condemnation. The other area of dispute involved the stingy Populist state legislature and the board of regents that they filled with their cronies and allies, who continually withheld financial support for the university. One example of his challenges occurred when the politicians fought funding for a new library building for KU by saying, "Lincoln had gotten along without one."

The breaking point for Snow's departure occurred with the sudden and tragic death of his son in the fall of 1899. The chancellor's son was working as a newspaper reporter in San Francisco when a large wave swept him off the deck of a small launch that was greeting a transport ship returning a Kansas regiment from the Spanish-American War. When Snow received the horrific news, he fell into deep despair. Never able to adequately recover from the death of his child and the stressful pressure of state politics, his depression and illness became unbearable. The chancellor took a long summer vacation in 1900 that turned into an extended leave of absence for the following school year. Finally, in the spring of 1901, when Snow's health was not showing signs of improvement, he was forced to relinquish the reins of the chancellorship and accept a faculty appointment as Professor of Organic Evolution, Systematic Entomology, and Meteorology. Such a designation appeared overblown considering that when he finally returned to teaching, he taught only a single class on evolution.

Because Snow had continued to serve as the director of the museum of natural history while chancellor, he had established a precedent for subsequent chancellors. That arrangement, though never specified by the university's charter or bylaws, worked well for Dyche and Bunk by giving them direct contact to the top person on campus whenever faculty infighting occurred. Although Bunk, as staff, never personally witnessed faculty jealousy, it was common knowledge that certain professors in the humanities departments believed that the natural sciences had always received favorable treatment at the hands of Chancellor Snow. Bunk couldn't dispute that assessment considering that during Snow's tenure as chancellor the university saw the construction of new buildings for chemistry (Bailey Hall) and physics (Old Blake Hall) compared to only the Spooner Library for the humanities. Now with Snow taking a diminished role, Bunk worried that the new chancellor, especially if not from the natural sciences, might lessen the museum's standing within the university community.

In spite of his understandable concern about the future of the museum, Bunk was also well aware that Kansas University had already made a significant commitment to bolster its standing. During Chancellor Snow's leave of absence, Professor Dyche and the university board of regents led the charge during the 1900 legislative session to convince the elected officials to fund the new building, and construction began the following year. Bunk regretted that he would not be a part of the opening of the new building, but a new building was not enough for him to pass up the opportunity in Oklahoma.

The timing of Bunk and Clara going to Oklahoma could have been better considering recent family events. A few months earlier, Bunk's mother had passed away just four days short of her sixty-seventh birthday. Bunk's father, who had recently retired from Haskell, moved in with Bunk's sister, who had recently relocated to Lawrence. Despite all the family upheaval, Bunk and Clara still chose to leave Lawrence, thinking that new opportunities must be taken when they present themselves.

The following events of Bunk's move to Oklahoma and his initial field trip are recorded in his contemporaneous field notes, written in 1901.

Dr. Van Vleet met Bunk's train when it arrived in Norman, around one o'clock in the morning. The professor informed Bunk that he would be spending the night at the Van Vleet home, which was welcome news, as he was exhausted from his long day. As they rode to the professor's house, Bunk learned that the departure date had been pushed back to Wednesday. This change allowed for another day to prepare for what was expected to be a two-month expedition to survey the animals of Oklahoma.

The next morning Van Vleet introduced Bunk to the other members of the collecting party. Paul J. White, a botanist, and Mark White, a general assistant and collector in botany, would collect plant specimens. The latter would also serve as the camp cook. Bunk and the professor intended to hunt animals, primarily birds. The four men spent Tuesday loading the wagon and readying the two horses that would pull the wagon. Early the following morning they clambered aboard the wagon eager to hunt the birds and collect the plants of Oklahoma. The first leg of the trip took them east of Norman. For forty miles they plodded through heavily forested hill country. Bunk wrote in his field notes that the "Jack Oaks were so thick in places it is hard to get a covered wagon through."

Their daily routine involved traveling to the next collecting site during the morning hours. In the afternoons, after they set up camp, Bunk and Van Vleet would hunt birds and Paul White would look for plant life, while Mark prepared supper. Bunk began making a list to record all the birds they either observed or shot. After a day and a half, Bunk and Van Vleet had about twenty different species recorded. That night they heard a call that Bunk thought sounded like a whippoorwill's, but without the "whip." Several days later, while Bunk was hunting around dusk, he shot a pair of birds. As he collected them, he solved the mystery of the strange cry of the whippoorwill. The call was that of another nocturnal bird called a Chuck-will's-widow. According to Bunk's field notes, he collected a nice variety of birds during that first week, including fine specimens of a Mississippi kite and a barred owl.

After a week on the road they circled back to Norman to drop off all the collected specimens at the university. Dr. Van Vleet also wanted to see if the check had arrived from the university to cover the expenses of their expedition. Assuming the money would eventually arrive, he announced that they would continue their trip. They spent the night in Norman before heading fifty miles north to Guthrie, intending to hunt around there for a few days before moving on to an area called the Salt Plains of Oklahoma, about a hundred miles northwest. When the wagon of men arrived, they were impressed by its unusual, rugged beauty. Bunk described it as "bottom lands below some very high Gypsum bluffs (two hundred and fifty feet high). There are streams running out of several canyons that are very salty in taste" that he estimated at "about 33 percent salt." He noted, "There are several small (manufacturing) plants located in the canyons evaporating the water and making very nice coarse salt."

From the salt plains they slowly worked their way south until they reached the small town of Watonga on July 5. Back in April, Van Vleet had been interviewed by the Guthrie newspaper. He explained that their survey trip would occur between June 10 and August 15. However, when Van Vleet walked out of the local post office, he informed the men that the check he had hoped to receive from the university to finance the remainder of their collecting trip still had not arrived. With only sixteen cents between the men and the party out of provisions, Van Vleet announced that he was suspending the survey. Before they drove back to Norman, Bunk got an idea that he included in his field notes:

> I thought I would get the start of [making a good impression with] the Dr. by sending a telegram to my brother-in-law for some money. I went to the station and wrote a telegram and handed it to the operator. He asked if I wanted to pay it or send it collect. I replied I wished to send it collect. He said I would have to put up a guarantee as that was a railroad line only and not a Western Union. I tried to talk him out of that notion as I did not have the price. But he was firm and I lost out and went back to the wagon. I said nothing.

For the next two years, Bunk and Van Vleet made numerous hunting trips around the Oklahoma Territory and the collection grew at a respectable rate. Unfortunately, the success of the growing collection abruptly ended in January of 1903, when a fire swept through the university science building and destroyed much of the collection. Gone were numerous field notes, 274 skins, and 92 of the 163 mounted bird specimens. Following that catastrophe, Bunk and Van Vleet attempted to rekindle interest in the collection from the powers that be, but the momentum for building a reputable collection of birds and mammals had waned. In hindsight, building such a collection may never have had much of a chance, considering how the territorial legislature initially underfunded the endeavor. From the very beginning the annual allotment was only $300 (less than $9,000 today) — barely enough to buy the first horses and wagon for collecting trips, much less sustain the program.

In June of 1903, Clara gave birth to the couple's first child, a daughter named Fedalma. The unusual name came from a character in George Eliot's poem *The Spanish Gypsy*. Soon after Fedalma was born, Van Vleet informed Bunk that the efforts to rebuild the collection were put on hold, so there was no need of his services. Greatly disappointed about how matters had turned out in Oklahoma, Bunk and Clara made the decision to return to Lawrence in the autumn of 1904. Bunk hoped that he could be hired on again with KU, and Clara wanted to be nearer her family, especially with the new baby.

The day after Bunk arrived back in Lawrence, he called on Dyche to see if he could get his old job back. Once he reached the top of Mount Oread, he saw up close for the very first time the newly constructed natural history museum. It had been completed and opened in 1903 and appropriately named Dyche Hall, to honor his former boss. Bunk was awestruck by the massive, beautiful structure. He thought that the size and architectural detail of the building offered strong evidence as to the importance of the natural sciences at the university.

Inside Dyche Hall, Bunk finally located the professor working at his desk in his office. He knocked on the frame of the open door and Dyche looked up to greet him. After they exchanged pleasantries, Bunk recounted the story of how most of his collection of birds and mammals at the University of Oklahoma had been consumed by a campus fire. When he asked if his old job was available, Dyche explained that he had no work to offer, but he would keep Bunk in mind should matters change. Bunk asked the professor about Williston, who he had heard left KU in 1902 to take a job at the University of Chicago. Dyche told him of Williston's concerns about the status of the natural sciences at the university after Snow stepped down as chancellor and director of the natural history museum. In addition, Williston had been discouraged about the death grip that the miserly state legislature had on the salaries at the university. Bunk was pleased to hear that Professor McClung had been promoted in Williston's place as head of the zoology department and curator of the paleontology collection, and that his good friend H. T. Martin was still working as McClung's assistant.

Dyche then offered to give Bunk a tour of his new pride and joy, the expansive exhibit space, which, at 72 feet deep, 132 feet wide, and over 20 feet high, was large enough to display all the mounted animals from the 1893 World's Fair. As the professor led him into the large room that occupied most of the main floor of Dyche Hall, Bunk's jaw dropped in amazement. Although it was in the early stages of construction, Bunk could tell that the new exhibit would be extraordinary. Dyche wanted his exhibit to be the largest of its kind in the world, in terms of both size and number of species represented. He proudly announced that the new exhibit would be named the *Panorama of North American Mammals.*

A year after Dyche Hall had opened, the exhibition area still looked like a construction zone. Lumber, timbers, and various building supplies were strewn across the floor, amid hammers, hand drills, sawhorses, and handsaws, all reflecting the work in progress on the diorama. Along the back wall of the exhibition space loomed two giant piles of wooden crates with mountain goats and bighorn sheep perched on top awaiting the finishing of the landscape. Amid the construction mess Bunk observed moose and

Panorama under construction circa 1903 in Dyche Hall (University of Kansas Natural History Museum records, University Archives, RG 33/0 Photographs, Kenneth Spencer Research Library, University of Kansas Libraries)

mule deer, which Dyche explained were ready to be installed. Other specimens destined for the display were still in the museum workshop being repaired or remounted. Although it had been only a little over a decade since the Chicago exhibition, many of the mounts had been damaged by frequent handling and exposure to the elements during their trip.

As Dyche explained his vision for the new exhibit, the progress already made impressed Bunk. He could see that the expansive diorama of animals would spread across the entire back wall of the large exhibition hall. All animals would be grouped in their own natural environment. For example, the bison and prairie dogs would be set in a prairie scene, the moose and mule deer would occupy a north woods landscape, and a mountain landscape from the Chicago World's Fair had been reconstructed for the bighorn sheep. To help create the illusion of standing in nature, a curved cyclorama had been erected inside the building's walls and wrapped around the animals. After the tour ended, Bunk left the museum more disappointed than ever that there was no job for him.

Once he exited Dyche Hall, he spent a few minutes walking around the

Dyche Hall, home of the University of Kansas Museum of Natural History (University of Kansas Natural History Museum records, University Archives, RG 0/22/17 Photographs, Kenneth Spencer Research Library, University of Kansas Libraries)

building to get a better look at the museum's exterior. The building was constructed in the Romanesque architectural style that was popular for public buildings and churches in the United States in the last half of the nineteenth century because it offered a sense of permanence and strength. The exterior walls were rough-hewn blocks of native limestone and topped with twin-hipped roofs adorned with red tile that flanked a flat roof that accommodated a skylight for the panorama. The ornate façade included stone archways, exterior decorative columns, and a sort of herringbone pattern border of red terra cotta and limestone just below the windows of the top floor. The octagonal wing on the rear of the building that would house the new panorama seemed in keeping with how Dyche revered nature. It made perfect sense to Bunk when the professor had told him how the building was designed to resemble a great cathedral with an apse for the panorama and a soaring square turret to mimic a church bell tower. To Bunk it looked as if Dyche Hall had been plucked from the streets of medieval Europe and set down in Kansas.

As Bunk's gaze zoomed in on the architectural detail of the building, he was surprised by one of the more unusual features of Dyche Hall. Positioned several stories above the ground were hand-carved limestone statues, or grotesques, crouching on small ledges. Each figure uniquely incorporated disparate parts of unrelated animals. Bunk felt that those mythical and monstrous figures were simultaneously both bizarre and whimsical. As described by one sculptor hired in 1996 to restore several of the grotesques, "one has the snout of a cow, the skull of a cat, (and) human breasts. . . . On this side it has split hooves on its back foot and on this side it has toes. Each one is a fanciful combination of animals both real and imagined." To the ever-practical Bunk, the juxtaposition of the strange statues seemed out of place on such a dignified building that represented a serious institution. Regardless of their propriety, Bunk found the grotesques captivating.

The decorative grotesques appeared as if standing guard over the names of famous natural scientists that were carved on the outside of Dyche Hall. The names AUDUBON (John James, 1785–1851, American ornithologist, naturalist, and painter) and GRAY (Asa, 1810–1888, noted botanist) adorned the north wall. COPE (Edward Drinker, 1840–1897, American paleontologist and comparative anatomist, herpetologist, and ichthyologist) and AGASSIZ (Louis, 1807–1873, Swiss-American biologist and geologist) were displayed on the south wall. On the front of building, in the most prominent position, were the names DARWIN (Charles, 1809–1882, English naturalist, geologist, and biologist, and best known for his contributions to the theory of evolution) and HUXLEY (Thomas Henry, 1825–1885, English biologist, often referred to as "Darwin's bulldog" because of his devoted support of Darwin and his theory of evolution). Although Bunk recognized all of the names, he wondered why those particular scientists had been chosen. He could only imagine how entertaining, and possibly heated, those faculty discussions could have been when they decided whom to include and whom to omit. It appeared logical to him that those selected represented a broad sampling of nineteenth-century scientists from various disciplines. However, when he noticed that Cope had been selected to represent paleontology, instead of the equally notable O. C. Marsh, Bunk could not resist a smile. He was fairly certain that Professor Williston had lobbied hard for that choice because of his earlier feuds with Marsh.

For the next year Bunk often stopped by the museum to visit with Dyche and the staff just in case an opening occurred. To make ends meet Bunk cut and sold firewood from the family farm and worked odd jobs. In addition, Bunk became acquainted with Charles H. Sternberg, who returned to his home in Lawrence after spending summers fossil hunting for Cope. As a result of that contact, Sternberg hired Bunk to prepare the fossilized skull of a small specimen of a mosasaur, a prehistoric aquatic reptile from the Cretaceous period.

CHAPTER SIX

Coming of Age as a Museum Man

During Bunk's hiatus in Oklahoma, the regents hired Frank Strong to succeed Snow as chancellor in 1902. Strong was a tall, slender man with close-set eyes and a mustache. He had originally come from the state of New York and later studied history and law at Yale. His early career included practicing law, working as a public school superintendent, and, just prior to coming to KU, serving as the president of the University of Oregon. Although Dyche Hall was not yet completed and open to the public, the formal ceremony for Strong's installation as chancellor was held in the cavernous room that would eventually become the primary exhibition space, because it was the largest space on campus.

Early in his administration, Chancellor Strong faced the decision of who should take control of the KU Natural History Museum. Both Snow and Dyche wanted to be the director. Despite decades of experience, Snow's mental health showed so little improvement by 1903 that Strong did not feel comfortable returning the mantle of director to him. Dyche had performed admirably in that role during Snow's absence but putting Dyche over his old boss might be awkward, especially considering that the two men had recently had a falling-out. Snow blamed Dyche for aggressive publicity surrounding credit for an achievement that Snow felt was his. To resolve the conflict, in February 1903 Strong announced that he would retain the title of ex-officio director of the museum, but appointed Dyche as curator of birds, mammals, and fish and gave him "virtual charge of all the Museum Building except the upper floor." Clarence McClung, as head of the department of zoology and curator of the paleontological collections, was given control of his part of the upper floor and Snow was named curator of the entomological collections (also on the upper floor).

After that decision was made, Snow's mental health began to steadily improve, to the point that he conducted ten entomology expeditions from 1902 to 1907. In addition, in 1903 Snow published his fifth edition of his reference book *Birds of Kansas* (the first edition had been published in 1872). Although insects were his primary interest, since 1866 he had always actively contributed to the bird collection. Despite Snow's diminished role, he admitted that he was enjoying his life away from the pressures of administrative duties. At a meeting of the Kansas Academy of Science in 1906, Snow declared that "in my room on the third floor of the Museum building, I live with my bugs, and beetles, and bees, and wasps and birds. I am perfectly content and feel as though already in heaven while still here on earth."

At the age of thirty-four, Bunk had been out of steady work for nearly two years when he received a telephone call saying that Professor Dyche wanted to see him. It was a hot August morning in 1905 when he climbed the familiar hill on Adams Street. After crossing Oread Avenue, he anxiously ran up the outside stairs of the museum building two at a time. He scurried through the heavy double doors and saw his old boss in the center of the expansive panorama exhibit space. The professor looked up from a long worktable covered with large sheets of the design plans for his panorama. He greeted Bunk and told him that he had a job for him to work on the landscape for new exhibit. Without hesitation Bunk accepted the offer.

Without knowing it, Bunk was taking the spot on the staff formerly held by his old friend Pug Saunders. During the time when Bunk was in Oklahoma, Saunders had spent most of his time restoring and refurbishing the mounted mammals from the World's Fair before they were to be installed in the exhibit space in the new museum building. As he completed the restoration project for the panorama, he suspected that he might be working himself out of a job. With few new animals to mount, the need for taxidermy was declining, at least until a very unusual project was offered to the university. According to an article in the *Lawrence Daily World* dated February 10, 1903, Saunders was preparing the skin of an elephant named Albert. Lemon Brothers Circus owned Albert and were performing in Kansas City's Argentine neighborhood when the elephant suddenly died. Saunders told the newspaper that "it would take at least a year before the skin, which weighs 1,000 pounds, is ready to mount." Today the museum has four elephant feet and some bones, but no record of the skin. Apparently, Saunders was less successful with salvaging Albert's skin than he was with the skin of Comanche. Soon after finishing the Albert project, Saunders left KU. For several months starting in April 1905 he worked for the Carnegie Biological Institute of Arts and Sciences at Cold Springs Harbor, Long Island, New

J. Charles "Pug" Saunders (Thomas L. "Tommy" Burns Collection, Kansas Collection, RH MS1907.8, Kenneth Spencer Research Library, University of Kansas Libraries)

York. Afterward he returned to Lawrence to make a living hunting and fishing, while working at a lumber planing mill.

As Bunk shook hands with Dyche to seal the deal, Bunk agreed to start immediately. Dyche led him over to introduce him to the other men working there that morning. As they walked toward the construction area, Bunk remembered a familiar face from his previous stint at KU: Leverett Allen Adams. He was a graduate student from Lawrence and would be leaving KU in a year after finishing up his master's degree in zoology. Next, Dyche introduced Bunk to a fresh-faced nineteen-year-old staff assistant named Theodore Rocklund, who had grown up in Lawrence and enrolled at KU in 1903. His father had emigrated from Sweden and supported his family of five by working as a tailor. Rocklund supported himself in college by working in the museum. Eventually he dropped out of college but continued to work in the museum because he, like Bunk, was more interested in the objective or curatorial aspect of the museum than the academic. He enjoyed taxidermy so much that one student even referred to him as "Theodore, the boy taxidermist." Despite their fifteen-year age difference, Bunk and Rocklund soon discovered their mutual love of hunting and became frequent hunting companions. Bunk and Theo, as Bunk always called him, developed over time a camaraderie and lasting friendship that extended to their families.

Around six o'clock, after Bunk and the others had worked on some of the infrastructure of the foreground of the diorama, Dyche suggested that they call it a day. As Bunk walked home back down the hill, he reflected on how happy he was to be back at the KU museum. Those years in Oklahoma and his time spent in Lawrence before being rehired at the university seemed like "lost years." The Oklahoma position never lived up to his expectations, and it had taken a considerable time to get back on his feet even after he returned to Lawrence. But that experience made him appreciate all the more the value of working at KU. The time he had spent in Oklahoma reminded Bunk of the old adage that some changes are not for the better. As a result, when he subsequently received an offer to be a preparator in paleontology at the Field Museum in Chicago and later the job of assistant superintendent of the Philadelphia Zoological Gardens, he politely and promptly turned them down.

Later that fall the last member to join the panorama construction team enrolled as a KU freshman. Alexander Wetmore graduated from high school in Independence, Kansas, where he had been living with his mother's brother. He had grown up in Wisconsin and, like Bunk, had developed a keen interest in nature at an early age. However, his father was a doctor and constantly nagged Alexander to give up his fascination with birds, calling it a "hobby." That conflict with his father eventually led Wetmore to move to Kansas.

The first meeting between Bunk and Wetmore occurred shortly after Wetmore enrolled in classes that fall semester. Because of Wetmore's interest in birds, he soon sought out the museum workshop. In some notes that Bunk prepared for Wetmore's birthday celebration in 1927, he wrote,

> I remember well the first time I ever saw him, a tall slim black-haired boy learning over a rail that we kept accross [*sic*] the door way in those days, to keep out visitors. He appeared there day after day looking longingly into the forbidden work shop. There came a day when the boss was away from town and I invited him into the shop and we became aquainted [*sic*] and friends. We made hunting trips together. I worked him into the shop first as a privileged character, later as a helper, and finally he was one of us.

Bunk quickly recognized that he and Alexander Wetmore were kindred spirits when it came to birds. "Doc," a nickname Wetmore received shortly after he arrived at KU even though he was still an undergraduate, was highly skilled as a hunter and already possessed an enormous knowledge of birds. Wetmore exhibited such skill and efficiency in skinning birds that Bunk also wrote in those 1927 notes about one occasion when Wetmore skinned eighty-five birds, recorded their measurements, made the catalog entries, tagged the specimens, and attended two classes of an hour each, all in eight hours. When Wetmore complained about not feeling well that day, Bunk wrote with a sense of amusement, "If he had felt well, I don't know how many he might have skinned."

Because of Wetmore's financial situation, the need to earn money to pay for his college education interrupted and extended his schooling. The uncle he lived with during his final year in high school worked for the railroad, and arranged jobs for the young Wetmore so that he could go to school one semester and work on the railroad during the next. For several years he worked in railroad stations in Arizona and California. Taking his experience and what he had learned at Kansas University, Wetmore collected birds when he was away from Lawrence. As a result, by the time he finished his degree at the University of Kansas in 1912, his personal collection consisted of nearly four thousand bird skins. He left his collection at KU with Bunk for safekeeping, although it was not officially part of the KU collection.

In October of 1905, just as Bunk was settling in with his new job, his father fell ill and died. Bunk had always looked up to his father and the simple life he had chosen for Bunk and his siblings. His passing reminded Bunk that it was important to do what you enjoy, regardless of how much money you earn. Bunk was not one to get emotional, so he remained stoic throughout the period of mourning. When he returned to the museum, it was business as usual, immersing himself in his work and the new panorama.

After Bunk rejoined the museum staff in 1905, the immediate focus of the small crew working on the panorama was to restore the 1893 exhibit in new exhibit space in Dyche Hall. Because Dyche, Adams, and Wetmore had responsibilities for classes and studying, most of the landscape construction for the panorama fell on Bunk and Theo. Although the infrastructure of wooden crates had been assembled to hold the mountain goats and bighorn sheep when Dyche had given Bunk his first tour of the exhibit in 1903, they now needed to turn the stack of boxes into a realistic facsimile of a mountain with craggy cliffs. To do that, Bunk and Theo began to transform the rough structure into lifelike rocks. Although Bunk and Theo had no formal training for this kind of work, Bunk had a keen eye for nature. In addition,

this early effort involved considerable trial and error. First they covered the wooden frame with wire mesh. Then they applied papier-mâché and plaster, sometimes with sand and straw added to the slurry to provide texture. Following that, they painted the surface of the rocks, bringing the ground work to life.

When the mountain landscape was completed, the animals had to be attached to the scene. This was made more difficult because the animals, which all came from the 1893 exhibit in Chicago, were never expected to stand on a flat floor; rather, they were mounted to fit the contours of that earlier landscape. That was intentional, so they would look more natural. However, that also meant that the new scenery in the panorama had to be built to match the preexisting pose of each animal. Once the mountain goats and bighorn sheep matched their rocky perches, Bunk and Theo needed a way to make the mounts stable. For that to happen, they developed some hidden fasteners on the animal to clamp a small rod attached to and sticking out of the landscape, effectively making the animals permanently part of the scene.

With the same high standards that Dyche insisted on for the taxidermy, Bunk and Theo were dedicated to creating an environment for the animals that looked as real as possible. Once the animals were in place, they began to dress the set with indigenous flora: bushes, grasses, lichen, and moss that Dyche had the forethought to collect from his hunting trips. If they did not have enough real fauna, they had to improvise and re-create the vegetation. The decision was made to set the scene in the panorama as an autumn day, which dictated that the grasses be brown and the leaves on the trees used in the panorama exhibit their fall colors. According to the 1921 University of Kansas Jayhawker yearbook, every leaf of the deciduous trees was real and

Panoramic photograph taken in 1906 of the *Panorama of North American Mammals* in Dyche Hall, University of Kansas (University of Kansas Natural History Museum records, University Archives, RG 0/22/17 Photographs, Kenneth Spencer Research Library, University of Kansas Libraries, courtesy of Karen Pendleton, high-resolution copy courtesy of Forcade Associates)

had "been soaked in a preserving solution, painted, paraffined, then wired and sewed to the trees."

Although the area set aside for the panorama was quite large, there were so many specimens placed in the exhibit that the scene might have seemed crowded compared to their natural environment had it not been for the skill of the men working on the landscape. For instance, with painstaking care they created a smooth, visual transition from the mountain cliffs for the Rocky Mountain goats to the prairie for the coyotes in only a few feet, whereas in real life that would have been miles, or hundreds of miles. Regardless, each individual tableau within the panorama was so convincing that the viewers were able to suspend their disbelief caused by the unrealistic closeness of the animals. As a result of this attention to detail, by 1906 the panorama of 1893 was restored to its original glory. To commemorate that, Dyche arranged for a local photographer, David M. Horkman, to take a panoramic photograph of the entire exhibit. Horkman took multiple photographs and matched up the edges to create an extraordinarily expansive picture of the diorama.

Dyche always wanted to improve and expand the panorama. Once the Chicago exhibit was replicated, he asked Bunk and Theo to add two-story cliffs to each end of the panorama. These would be located at the two corners formed where the main section of the museum connects to the rear addition. The granite cliffs on the north end of the exhibit would showcase the forest animals. The ever-resourceful Bunk and Theo pulled old crates and barrels out of storage and recycled the wood to construct the framework. Some of those containers had previously been used to ship the mounted animals and materials to and from Chicago for the fair, as well as earlier hunting expeditions. They even commandeered a wooden box that had been used to send books to KU from the West Publishing Company (today the publisher's imprint is still visible from behind the scenery).

On the south wall of the exhibit Bunk and Theo constructed a sandstone outcropping to highlight the prairie scene of coyotes and bison. This infrastructure required more than recycled crates, so they sought out longer and continuous lengths of lumber. Although neither had studied the formal principles of structural engineering, the underpinnings of those important features were strong enough to support the weight of the landscape for over a century.

Even though the work on the panorama filled Bunk with pride, he never sought any special recognition. However, his efforts were acknowledged by a later director of the museum in a biography of Francis Snow, which said that Bunk was "skillful in erecting dioramas." As for his friend and accomplice, the impish Theo took time to paint his name in large letters on the back side of cross members of the scaffolding that supports the section depicting the Rocky Mountains. His signature remains today, though out of the public eye.

After returning to KU in 1905, when Bunk wasn't working on the panorama, he became increasingly involved with the field and workshop training of the students. Bunk was thirty-four years old at that time, which typically made him at least ten years older than the college students. That age difference provided Bunk with the credibility of experience. Also, by that time he had earned a reputation as a seasoned taxidermist, possessed an extensive knowledge of birds and animals, and was considered a skilled field collector. Although there was no formal line of authority between Bunk and the young students, they began to gravitate to him for advice and counsel.

Bunk enjoyed teaching in a one-on-one setting, a talent that he first observed in the workshop with Williston. Because of Bunk's humble and unassuming personality, the students immediately felt comfortable with him. Bunk never intimidated the students when they lacked certain scientific knowledge or needed help with any aspect of the collection process. His patient and genuine concern for their education put them at ease. But despite

Bunker (*right*) and Theodore Rocklund constructing limestone cliffs on the south wall of the panorama circa 1907 (University of Kansas Natural History Museum records, University Archives, RG 33/0 Photographs, Kenneth Spencer Research Library, University of Kansas Libraries)

his easygoing manner, he held all the students accountable for doing the work correctly and accurately. As a result, the students increased their self-confidence as well as their scientific knowledge.

Bunk firmly supported the "direct study of nature" philosophy — he believed that the students learned more when they participated in collecting specimens. Observing an animal in its natural environment offered more information about the creature than some inanimate object in a collection drawer. Bunk's classroom of choice was a farm located seven and a half miles southwest of Lawrence. Giving the pretext that he needed to cut firewood, most weekends Bunk invited students to a day of hunting. But the truth of the matter was that Bunk thought of those outings as educational opportunities. So as not to waste valuable daylight, Bunk and the students met at sunrise at the "animal house," the name given to a barn on the southern slope of Mount Oread that housed all the university's livestock and test animals. They carefully gathered everything they would need for the day: guns, ammunition, usually a little something to eat for their noon dinner, and canteens of water. In the early days they traveled to the farm in a horse-drawn wagon, but the advent of automobiles shortened the trip and allowed more time for hunting. Every time Bunk and the hunters went to the farm, they drove south out of Lawrence on a dusty dirt road, today known as US Highway 59 or south Iowa Street. They would talk, tell stories, and try to identify birds along the way to make the time go faster. The road took them past fields of wheat, corn, and beans. Once they crossed the bridge at the Wakarusa River, they turned right on the first road and continued for nearly a mile before turning south toward wooded hills that bordered the floodplain. Their destination was a small rustic one-room cabin nestled in a narrow, heavily wooded valley. It was surrounded by hills, streams, and cropland that provided an abundance of wildlife for the students to hunt. Almost any weekend throughout the year, Bunk and the students hunted this area known to the locals as "Washington Creek."

Rough-hewed planks of wood covered with tar paper formed the shell of the rustic cabin. A pitched roof ran the length of the structure, a row of windows on one side provided light, and a small covered porch on the narrow front of the shack protected the door and stoop from the elements. With no running water, plumbing, or electricity, it certainly qualified as primitive. Despite the lack of luxuries, the students lucky enough to hunt there always remembered the old cabin fondly. Whenever they gathered in later years, they always recalled the camaraderie, being out of doors, and their time hunting. Over time the cabin became known affectionately and simply as "Bunk's cabin."

The first order of business upon arriving at the cabin involved Bunk telling the first-timers the lay of the land: where they had permission to hunt, where the best hunting grounds were, and what time they would start back to town. Afterward, the students headed out looking for game. Whenever the students shot a bird or a mammal, they brought it back to the cabin where Bunk was collecting firewood. Bunk always made certain that the cabin contained knives, tweezers, and chemicals for preserving the skins — all necessary for specimen collecting. As usual, Bunk offered assistance, if needed, and made sure all steps of the collection process were followed. He required the students to correctly identify the animal, always insisting that they use the scientific name. Next, he made sure they recorded the specimen in their field notes, field dressed the animal, and carefully tagged and stored it for return to the museum.

With Bunk's oversight, the students learned "hands-on" the entire process of adding a specimen to the museum collection. Typically Bunk stood back and let the students go through all the steps in the process, knowing

Bunk's cabin south of Lawrence (family photo)

that if they did everything on their own they would learn more. According to one student who later wrote a memorial for Bunk, even though Bunk could always identify the species, he motivated the students by telling them that his "lack of education and scientific know-how required him to depend entirely on the student to insure that all was in order and scientifically accurate." Such mild-mannered and well-meaning deception reinforced the idea that the student's work was important and essential. Over time the students became indebted to Bunk for the trust and confidence he placed in them. Bunk continuously motivated the students with encouragement, praising them for every specimen they brought to him. Bunk also made the students feel special because each specimen they gathered "was an early record, or a late record, or it was needed to fill out a series." In addition, Bunk made it clear to the students that the university collection must grow to justify its existence, and he was counting on their help to grow it. As a result, the students felt like they were keeping Bunk in good standing with the university, and one of his students wrote that because of this, he "worked harder than [he] would have otherwise."

Bunk always made certain that the students got credit for the specimens they added to the collection because he understood how much recognition and being appreciated motivated them. Bunk had a way of making suggestions to the students, rather than giving instructions. If their inexperience meant they required more time to complete the process, the students always appreciated Bunk's patience. Bunk knew that learning to correctly add a specimen to the collection would pay dividends later for both the student and the museum. However, given autonomy and responsibility, the students occasionally damaged a delicate skin during the preparation, which could lessen the quality of the specimen or render it useless for the collection. But that never bothered Bunk, because he cared more about the educational experience for the students.

Although Dyche spent considerable time with the students, Bunk's time with them created a strong and lasting bond, a kind of relationship that never existed for Dyche. Dyche was well respected for his intellect, speaking skills, and vision, but the students saw him as a person more focused on the powers that be and the politics within the university, the state government, and the public at large. In Bunk, on the other hand, they saw an unsophisticated, working-class man with limited social skills, but someone so sincere and caring that they referred to him as "salt of the earth." Consequently, they admired and respected Bunk, much like how the paleontology students felt about Williston.

The weekend ritual of hunting at Bunk's cabin that began in 1905 continued for decades. Besides the invaluable educational experience, the students enjoyed the companionship of the other boys and built lifelong friendships. In addition, those weekend outings demonstrated to Bunk which students possessed promise and aptitude for life as a naturalist. Every new generation of students that joined those collecting trips to the farm understood that they were so fortunate to be part of an exceptionally elite fraternity that eventually came to be known as "Bunk's boys."

One day in 1907 Bunk came across an article in Ridgway's 1901 edition of *Birds of North and Middle America* about the black-capped vireo, a small songbird that is found today in dense thickets of brush in the hot canyons of Texas, Oklahoma, and Mexico. Ridgway's reference book was admired and widely considered reliable, but Bunk took issue with the findings of the article. The article concluded that the color of the plumage of the blackcaps did not distinguish or correlate with the gender of the bird, which was different from Bunk's own personal experience. Years earlier, while doing fieldwork for the University of Oklahoma in Blaine County, Oklahoma, he observed black-caps for three weeks during their breeding time. Believing that the reference book was incorrect, he shared his concern with Snow, the most prominent ornithologist at the museum. After their conversation, Bunk contacted Professor Henry Higgins Lane, the zoologist at the University of Oklahoma, to ask if he would send the series of black-cap specimen skins that Bunk had collected back in 1901 and 1902. When the specimens arrived, Bunk showed them to Snow, who agreed with his assessment. The professor encouraged Bunk to submit an article to the Cooper Ornithological Club, one of the country's oldest and most prestigious organizations for scientists studying birds. But Bunk lacked confidence in his writing abilities and put off starting the article.

By 1909 Snow had been diagnosed with a coronary disease. One autumn morning he awoke and announced that he felt "different." Without notice, he immediately collapsed and died peacefully. Hearing the news saddened Bunk. Snow had always been so kind and encouraging. Bunk also regretted that he had not begun the work on the black-capped vireo article, especially since Snow would not be around to advise him. A month later, to honor the late professor, he commenced writing his article with help and advice from Dyche, and later that year he submitted it for publication.

In March of 1910, Bunk received word that his article, "Habits of the Black-Capt Vireo (*Vireo atricapillus*)," was being published in the *Condor*, a highly respected "magazine of western ornithology" published by the Cooper Ornithological Club (which merged with the American Ornithologists' Union in 2016). Bunk had laid out in a logical and comprehensive manner his argument for how plumage correlated with the gender of the birds. His first real scientific publication delighted him; it was a feeling he would always remember, and it was one reason he always encouraged his students to publish, so they could experience that same sense of pride.

By 1909 the installation of large panes of glass finally protected the *Panorama of North American Mammals*. Dyche vowed that the panorama would never be entirely completed, because it would continually evolve and change with new additions. But with most of the animals in place and the majority of the landscape work completed, when Kansas governor Walter Stubbs approached him about taking the position of the state fish and game warden, Dyche gave the invitation serious consideration. Heading up the office responsible for protecting Kansas wildlife interested him because of his concerns about how humans endanger species. Kansas laws controlling hunting limits on game animals had existed for years, but a long-standing custom by local game wardens to disregard the enforcement of those same laws thwarted state policy. That practice persisted due to political patronage, where local officials appreciated the rewards and prestige of those positions more than the unpleasantness of enforcing the game laws upon their neighbors. The governor hoped that Dyche, as a scientist and vocal advocate for protecting nature, would bring professional credibility to the office and change that culture.

When Chancellor Strong learned of the governor's offer, he wrote a letter to Dyche encouraging him to accept. With his statewide reputation so closely tied to KU, Dyche's service in the high-profile position of game warden would no doubt enhance the image of the university. As it turned out, Dyche's decision revolved around his university pension. Since the position of game warden was an appointed one, he worried that a new governor might dismiss him, which would cause him to lose his university retirement benefits. Dyche negotiated a deal with the governor and chancellor to assume the state post while continuing to be an employee of the university as curator of recent vertebrates, thus preserving his university pension.

To oversee the day-to-day activities of the recent vertebrate collection while Dyche spent time at the state capitol in Topeka, Strong appointed Bunk "assistant curator in charge." With Bunk's new title came the responsibility to represent the museum to the public at large. Some of those new duties he relished and others caused him extreme discomfort. Bunk could talk one-on-one with students for hours about birds and mammals, but public speaking terrified him. Despite years of observing the always confident L. L. Dyche entertain audiences, none of the professor's panache had rubbed off on Bunk. Shortly after he became assistant curator in charge, he was asked to give a speech about birds to a local women's group. Even though he wrote out his remarks in advance, when the time came to speak he froze up. After that embarrassing experience he always avoided accepting future speaking engagements.

Although he sought to avoid public speaking about the museum and the collection, he was more than pleased to field individual scientific questions. The citizens of the state looked to the museum of natural history as a scientific resource. For instance, whenever they observed something they considered might be of scientific value, they often contacted the university to come and look at it and render an opinion. A prime example was the farmer who had notified the university about the fossilized bones of the Twelve Mile Creek bison. That kind of service to the state occurred with some regularity, enough that Theo kept a journal of those instances whenever someone local contacted the natural history museum about a dead bird or mammal that they had found. He would investigate the incident and note whether or not he brought it back for addition to the collection.

Besides representing the museum, Bunk served as the custodian of the collection of birds and mammals. He was responsible not only to care for the specimens but to make them accessible to faculty and students. By 1909 the collection had grown to the point that it became unwieldy and in need of some organization. Fortunately, Bunk was the right person for the job. Dyche had been an extraordinary force and possessed amazing vision, but he never relished routine. While it was the daring Dyche who scoured North America for large mammals, he had little interest in something as mundane as a cataloging system. Rather, he preferred keeping all the information about any given specimen in his head. But because subsequent scholars would not possess that same body of information, the museum needed an efficient way to locate the specimens they wanted to study.

Under Dyche, the system for storing specimens was exceedingly simple for the collector but ineffective for the researcher. It involved handwriting a number on a rough label that was attached to the skin or skeleton. These old specimen tags lacked uniformity and permanence, consisting of a variety of materials: usually wood, metal, cloth, or pasteboard. The number on the tag corresponded with the number found in one of Dyche's many field notes, rather than one centralized listing. As a result, it proved difficult for anyone but him to study the collection efficiently.

In 1910 Bunk began implementing a standardized system that identified all of the specimens in the collection. Given the number of existing specimens, it would take years to fully implement. To each specimen, an acid-free pasteboard tag inscribed with permanent ink was attached — materials that would last for years. Bunk also recognized the need for a central record of all specimens in the collection. Instead of relying on memory and assorted field notes as the primary source of scientific information, he created a centralized system that utilized large ledger books, also with acid-free pages and permanent ink. The collector would attach a temporary tag to the specimen and number, which would correspond with an entry of the same number in his field notes. When the specimens and field notes returned to the museum, every new specimen was assigned a unique catalog number in the order of when it was added to the collection. That specimen number from the catalog would be written on a new permanent tag on the specimen, which replaced the temporary tag.

Another advantage of Bunk's new catalog system was that now more detailed information could be maintained. Each entry in the large catalog included the proper scientific name, including order, family, tribe, genus, and species. In addition, the record included the location of the specimen's collection — for example, Lawrence, Douglas County, Kansas, seven and one half miles southwest of Lawrence (which, incidentally, was the way they always referred to Bunk's cabin). Finally, the date it was collected, the name of the collector, and the measurements of the specimen were noted. With Bunk's new catalog system there was no duplication of numbers, and it reduced reliance on the field notes. Bunk also anticipated that ornithologists and mammalogists might appreciate having their own separate catalogs, so he created separate ledgers for birds and mammals.

Bunker understood that future naturalists would require more order, especially since he knew the number of specimens in the collection would continue to grow. That was particularly important now, with all the talk around the museum of embarking on an extensive survey of Kansas vertebrates the following year. The young students who studied under Bunker learned the new cataloging system and understood its value, too. When they left the university, they took their knowledge of the system with them and implemented it wherever their careers took them. As a result, Bunk's system was adopted by a significant number of universities and museums, and it is still the basis of most catalog systems today. Of course, today's collection catalogs are computerized, allowing even more information to be captured — for instance, the exact latitude and longitude of the location of the collection, whether frozen tissue samples are available for a specific specimen, and its DNA.

In May of 1910, the specimen cataloging project was interrupted by the arrival of Bunk's second daughter, Audrey. The day of her birth saw Bunk pacing on the porch while Clara was tended to by the doctor and Clara's mother. When the doctor finally exited the front door to announce that mother and baby girl were doing well, Bunk grinned and lit up a cigar to celebrate. That was about all the emotion Bunk could show for one day. He returned to the museum early the following morning, of course with a handful of cigars for his colleagues. Clara did not begrudge his departure that morning; she recognized and appreciated Bunk's many talents, but she knew that helping with a newborn baby was simply outside his skill set.

On top of the pleasant addition of another daughter to his family, Bunk's last five years had been eventful and professionally rewarding. After his return to KU, he and Theo had spent years and untold hours bringing to life the *Panorama of North American Mammals* by carefully creating the detailed and lifelike landscape. His fieldwork at his cabin with the students enhanced his reputation as a field zoologist and marked the beginning of his long career of nurturing young naturalists. His promotion to assistant curator in charge when Dyche became the state game warden encouraged and inspired him to take on the responsibility of developing an effective and efficient catalog system to keep track of the growing multitude of museum specimens. His article in the *Condor* added another arrow in his quiver and broadened his résumé as a serious "museum man."

For a man initially hired as a taxidermist, the idea that the drastic reduction in taxidermy activity that had occurred after the turn of the century might affect his job security apparently never occurred to Bunk. Going back to 1895, Bunk observed that all job descriptions at the museum were loose approximations. From his time projecting photographs with the magic lantern for Dyche's lecture tour to building the panorama, Bunk evolved and adapted to the needs of the museum and the university, just like any creature in its environment in nature. The one overriding philosophy of the museum in those days empowered everyone to do whatever needed to be done. The pragmatic Bunk could not have asked for a better job.

CHAPTER SEVEN

Beginning the Long Journey West

When Professor Dyche served as the curator of birds and mammals at the University of Kansas, he believed that future historians would benefit from a record of his life, so he retained volumes of correspondence. In fact, he even hired a clipping service during his career to accumulate newspaper articles to preserve his accomplishments for posterity. Bunk, on the other hand, never considered his life to be that important, so the record of his life is meager. As evidence of that difference, today the University of Kansas Archives has eleven boxes of articles and correspondence dedicated to Dyche, while Bunk's solitary box is only half full, even though his role as curator spanned nearly thirty-five years.

Considering the general dearth of information relating to Bunk's life, it was auspicious that some of his old field notes stored in the walk-in vault on the top floor of Dyche Hall survived. They provide the only authentic and contemporaneous record of his 1911 and 1912 specimen-collecting trips to western Kansas. The expeditions greatly contributed to the museum, substantially increasing the bird and mammal collection as well as an unexpected discovery of one of the university's most significant fossil specimens. This is especially fortunate considering that the notes had gone missing for decades before resurfacing sometime in the 1990s. In the next four chapters, all accounts of places and events, and quoted material that occurred between May and October of 1911 and September and November of 1912, are taken directly from Bunk's field notes.

Spring and early summer thunderstorms commonly occur on the plains of Kansas, occasionally spawning powerful and destructive tornados. Random and unpredictable, those twisters wreak considerable havoc, property damage, and loss of life. Such was the case on April 12, 1911, when a tornado roared through Lawrence around eight o'clock on a Wednesday evening. Preceded by thunder, lighting, and hail, the tornado funnel dropped out of the clouds a few blocks north of the university. The twister then traveled northeast, destroying nearly a hundred homes and the north end of the downtown business district before crossing the Kaw River. Two people died as a result, and many others sustained injuries. Fortunately for Bunk, the south side of town and the university were spared.

The morning after the storm, Bunk walked downtown to view the damage before going to work at the museum. The business district was a disaster; roofs and top stories of many buildings had been blown off, electric lines lay precariously on the ground, and most electric power was off. Rather than walk back on Massachusetts Street and go up Adams Street to get to the museum that morning, he wanted to see the residential destruction. As he tracked the path of the twister, he soon arrived at the residence of Charles H. Sternberg and found the fossil collector clearing debris in his front yard. As the two men commiserated about the storm damage, Sternberg related that the fossils in his workshop had been spared, for the most part. Bunk shared with Sternberg his upcoming expedition to western Kansas to collect specimens of birds and mammals. He proudly explained that he was feverishly preparing to depart in just a few weeks for an extended trip.

Since 1866, in spite of the informal efforts of Frank Snow to survey all birds in the state of Kansas, the university had primarily justified its natural history collections as teaching tools. It wasn't until 1911 that the state of Kansas and KU formalized the value of a comprehensive record of the diversity of plants and animals of the state by establishing the Kansas Biological Survey (KBS). On May 10, 1911, Chancellor Strong petitioned the board of regents to fund the new survey at the University of Kansas. On June 16, 1911, Strong's request was granted and $500 was allocated to fund the new organization. As a result, nearly a dozen scientific expeditions left KU that year to survey the flora and fauna of the state. Those collection parties were to be under the direction of a board from the University of Kansas made up of the chancellor; the heads of the KU departments of zoology, botany, and entomology; and the state game warden (Dyche).

In anticipation of the creation of the KBS and the associated funding, the museum began making plans for a variety of extended collecting trips beginning that summer. With guidance from Dyche and McClung, Bunk began thinking about how to approach such a massive undertaking. He knew that the university's collection of Kansas bird and mammal specimens at the time primarily represented the animal life of the eastern half of state, so it was decided to survey the birds and mammals of western Kansas. Because of the time required to reach the western part of the state and to adequately survey such a sizable area, they began planning survey trips in both 1911 and 1912. Birds and mammals are more prevalent in the spring and fall and the dry summer was better for fossil digging, so the expedition for birds and mammals was coordinated with Handel Martin's summer fossil trips. Bunk wanted to leave Lawrence as soon as weather permitted, hopefully by the first of May. McClung recognized that a May departure date would likely precede the funding approval; however, the benefits of getting an early start on the expedition outweighed the risk of any out-of-pocket expenses that Bunk would incur should the regents deny the KBS funding.

Bunk wanted to depart before the end of spring classes, so he asked Theo Rocklund to join him on the long trip, since neither had teaching obligations. Bunk also chose Theo because of his experience and skill in hunting, field dressing, and camping. In addition, Theo was self-motivated and willing to do whatever was needed without having to be asked. But most importantly Bunk and Theo always enjoyed each other's company. Compatibility was essential given that the two men would spend months together.

As soon as McClung gave his approval, Bunk and Theo began planning the trip. In spite of the fact that automobiles and trucks had become more widely available by 1911, Bunk elected to travel by the university's horse-drawn wagon. An automobile or truck might have saved driving time between collecting fields and offered more protection and comfort from the weather, but Bunk and Theo were more familiar with the team and wagon and quite satisfied with that mode of transportation. Their rig was a planter wagon, very similar to a buckboard but with higher sides. At each of the four corners of the wagon, tall posts supported the hard-shell roof. Canvas curtains securely fastened to the roof on the back and both sides could be rolled down to protect the contents of the wagon from any inclement weather. Bunk and Theo, on the other hand, would ride up front on a hard bench seat, where the small overhanging roof would provide little if any protection from a heavy rainstorm.

For nearly two months Bunk and Theo gathered and cleaned the equipment and gear that they would need for the long journey. They arranged for a tent, a dining fly, cots, sleeping bags, and a couple of kerosene lanterns.

In addition, they accumulated canned food, kitchen utensils, and a collapsible cooking grate. For collecting birds, they cleaned their shotguns and assembled shell casings, gun powder, primers, and a small press to reload the shotgun shells. They planned to trap small mammals with an assortment of both wire and spring traps of various sizes. They also packed their rifles to shoot larger mammals and to protect themselves on the odd chance they encountered any predators in the wild that might try to "collect" them. They assembled all the necessary tools, equipment, and supplies to safeguard the collected skins: folding worktables, knives, tweezers, and chemicals to both preserve and euthanize animals. They packed several twenty-five-foot loops of twisted manila rope for general use and a still camera and film to chronicle their trip. Finally, they packed a box with a stack of topographic maps to help them find their way. Because topographic maps only cover an area roughly the size of a county, they had nearly one hundred maps in the map box. As their departure date approached, they spent hours packing and repacking all the equipment, making sure everything would fit in the wagon. Lastly, each man packed a small suitcase with a few clothes and toiletries, keeping in mind how cramped the wagon was and knowing that equipment and provisions had priority.

Completing the expeditionary team were Billy and Baldy, who would literally provide their horsepower for the next six months. Both Bunk and Theo were thoroughly familiar with the two horses: Billy, the more docile, could graze untethered without risk of running off. Baldy, on the other hand, was more adventurous and always needed to be tied up to keep him in camp. Those horses would be integral to the success of their trip; if anything happened to the horses, such as injury or loss, the expedition would come to a halt. A few days before they were to leave Lawrence, Theo saw to it that the winter coats of both horses were trimmed, and he had a blacksmith fit them with new shoes.

On May 1, 1911, the much-anticipated day of departure arrived. Bunk and Theo had agreed the night before to meet at seven thirty; however, both men, filled with excitement, arrived early. The low angle of the early morning sun cast long shadows as Theo waited for Bunk to unlock the double door of the animal house. By all accounts it looked as if their departure day would be sunny and crisp. Unlike the warm muggy weather that had generated the devastating tornado a few weeks earlier, the morning of their departure offered a stiff northerly breeze. Scattered clouds danced across

Theo and Bunk pretrip preparations, 1911 (University of Kansas Natural History Museum records, University Archives, RG 41/0 Photographs, Kenneth Spencer Research Library, University of Kansas Libraries)

the sky, occasionally allowing the sun to peek through. Although they had hoped for springlike warm weather, the north wind served as a harbinger of the weather they would encounter.

When the doors swung open the two men entered, taking no time to admire their efficiency in filling their wagon. Instead, they immediately began moving quickly around the barn. Theo hitched the horses to the wagon while Bunk looked over the contents of the wagon one last time to see if they had forgotten anything. With everything in place, the horses pulled the wagon out of the barn. Theo locked the doors to the animal house and joined Bunk on the seat of the wagon. At a few minutes before eight thirty in the morning, with a flip of the reins and a click of the tongue to signal the horses, the wagon creaked and jerked forward. By Bunk's own admission, they began their trip "with the most complete outfit for a collecting trip that ever left the University."

Bunk steered Billy and Baldy south out of Lawrence. As soon as they passed the city limits, the two men fell into a state of quiet reflection, as they slowly bounced down the road in silence. Both men watched the road and the gait of the horses for possible injury. They also listened to the calls of the birds along the way. Theo broke the quiet by raising the question of whether

Wagon and horse rig for 1911 trip to western Kansas (University of Kansas Natural History Museum records, University Archives, RG 33/0 Photographs, Kenneth Spencer Research Library, University of Kansas Libraries)

they would run out of subjects to talk about, considering that they would be spending so much time seated together on the bench of the wagon. Bunk assured his young partner that that would not be a problem. Even if they did, both men were accustomed to spending considerable time together without talking. While Theo was glibber and more talkative than Bunk, the two men enjoyed each other's company and never felt compelled to fill up the silence with unnecessary conversation. Bunk reminded Theo about the years and countless hours they had worked together building the panorama, when their conversations were little more than an occasional request for a tool or a helping hand. Also, they had hunted for years and seldom talked then — although that was mostly to avoid spooking their prey.

For the first seven miles they followed the same route that they always took to Bunk's hunting cabin, but when they reached the narrow dirt road leading south they continued on west toward the small farming community of Clinton. As they plodded along, Bunk began to worry whether they had packed everything. He went over the list again and again in his head. In spite of his earlier declaration of being well prepared, a shot of adrenaline hit him when he remembered that they had packed only one feed box for the horses. Fortunately, Clinton was just a few miles down the road and he knew the general store would sell him another.

By noon they stopped for dinner on the side of the road a few miles short of Clinton. The two men ate the sandwiches they had packed and let the horses graze. As a cold breeze continued to blow, Bunk and Theo began to question their decision not to be better prepared for cold weather. According to Bunk's field notes, they thought it would be warm for most of the trip, so they elected not to burden themselves with too much warm clothing and bedding, especially when the wagon was so full. For the time being, they settled for just hoping the weather would soon turn warm.

After dinner they drove the wagon to Clinton, which today sits underwater at the bottom of a Corps of Engineers lake by the same name. They stopped at the general store and bought the needed feed box, some large spoons, and several cans of sardines. From Clinton, they drove the wagon on to the home of a friend, Jonnie Mosses, who lived nearby in Twin Mound, an early townsite on the western edge of Douglas County. Bunk and Theo arrived at Mosses' place at about three thirty and spent the rest of the afternoon catching up with their old friend. Because the temperature was dropping, they were pleased and relieved when Mosses invited them to spend the night in his house.

Before going to bed, Bunk took out his notebooks to record the events of the day. On this trip Bunk had decided to keep three separate notebooks. The first was a simple diary, in which he recorded the date, where they spent the night, daily activity, road conditions, who they met, the daily total of animals collected, how the horses were faring, any reimbursable expenses, and occasional addresses of people they met. In a second notebook Bunk listed each bird and mammal skin collected. Each entry included the species, number, the location where the specimen was collected, and the length and the gender of the specimen. There was also a notation of whether the specimen was an adult or juvenile. The third notebook listed birds Bunk observed, but did not collect, and was written in a narrative fashion. For instance, on May 2, 1911, Bunk observed "one barn swallow, one purple martin, one sparrow hawk, a small flock of yellow headed blackbirds . . . several hundred red wings, male and female, but while they were in one bunch, the males were more or less by themselves in one tree and females by themselves in another tree."

The next morning Bunk and Theo were greeted with cold, cloudy weather. Because they had covered so little ground on the previous day, they got on the road early. At eight o'clock the road was dry and hard, and the going was relatively easy. By midmorning, they reached the coal country of Osage

County. The first town they came to was Carbondale (an obvious name for a coal town). The town had boomed in the 1870s when coal was discovered nearby. This area of Kansas boasted numerous coal companies with dozens of strip and shaft coal mines. At that time, coal was a great source of energy for Kansans. It could be used to make coal gas for lighting and could be burned to heat homes and businesses, but its greatest value was to the railroads, which annually consumed tens of thousands of tons of coal to run their locomotives. In the late 1880s the railroads' demand for coal exceeded the entire output of Osage County, causing the railroads to look for and find other sources. It turned out that those newly developed mines began to produce coal at a lower price than Osage County companies, so the boom years in Osage County soon evaporated.

Approaching the town of Carbondale, Bunk told Theo about how Barnum Brown, the young student and stellar fossil collector who had attended KU prior to Theo's arrival, came from the town. When they reached the center of town, Bunk was surprised to find the general store closed. The two men asked around and were told that there was a funeral in progress at the church and the store would open after the "funeral had passed." In his notes, Bunk joked that he thought "it must have been the mayor who had died." Not wanting to wait until the funeral ended, they continued on to the next town. By early afternoon the horses and wagon had traveled ten miles to the town of Scranton, named after the coal town in Pennsylvania. If Bunk thought that Scranton was going to be an improvement over Carbondale, he was sadly disappointed. Even though both Scranton and Carbondale fell on hard times with the decline of the coal business, apparently Scranton fell harder. According to local historical records, Scranton had some of the less seemly businesses associated with coal companies, like company stores and saloons. By 1911 closed stores and saloons, in addition to the abandoned and dilapidated homes of the laid-off workers, dotted the landscape of the small community. Bunk noted, "It was a town that did not look good to us." With that observation, they continued on without stopping.

As they drove on to the next town, Bunk reflected on the general appearance of those declining old coal towns. In some ways they reminded him of Mendota: places that once thrived and later succumbed to changing economic times. Both the coal towns and Mendota were situated in agricultural areas, but their economies relied heavily on industry, coal, and railroads, respectively.

As the afternoon shadows began to lengthen, Bunk and Theo reached

Burlingame, a town located where the Santa Fe Railroad crossed the old Santa Fe Trail. Like thousands of small agricultural communities across the state that catered primarily to residents of the surrounding area and not to travelers, Burlingame had no hotel. Without any public accommodations, Bunk and Theo would have to camp that night under the stars. They drove around the town for an hour looking for a place to set up camp, eventually finding a spot on a vacated road just outside of town. Since it looked like it might rain, they were careful to secure their tent and fly. They also tied down the side covers on the wagon. According to Bunk, they were "well fixed." The next day they planned to go into Burlingame to find a blacksmith, because Baldy had thrown a shoe. While they were in town they could replace Billy's collar, which was too tight.

With their first extended hunting ground weeks away in Gypsum Hills, in south-central Kansas, they planned to drive the horses from morning to sundown. Such a plan would likely put them at the end of the day someplace where there might not be a town, much less a hotel. Unlike the early days on the old Santa Fe Trail, when travelers could just camp wherever they wanted, by 1911 nearly all the land in Kansas was settled and in private ownership. Bunk and Theo knew they could camp overnight on private property only with landowner permission, or else they had to find public land. Fortunately, at that time public schools were widespread across the state of Kansas and available for travelers to camp. Regardless of whether school yards were in towns or in small unincorporated communities between those towns, they always were large enough to set up camp and provided a source of grass for the horses to "pick," which was what Bunk called grazing.

The next morning, as they had expected, it was raining. Rain fell throughout most of the night, and the clouds showed no sign of breaking up. The men pulled up stakes after breakfast and drove to the business district in Burlingame to take care of the horseshoe and collar. Leaving the blacksmith shop around ten thirty in the morning, they found the roads quite slippery. Continuing rain worried Bunk that they would cover few miles that day. Travel in 1911 depended mostly upon the weather and the soil conditions of the roads. When it was dry and not too sandy, the horses could travel thirty miles or more. If it was raining and the road turned to mud, they might make only five miles a day. Despite the road conditions, a comprehensive road system of section-line roads existed across the state to give farmers and ranchers access to the property. For most of their journey Bunk and Theo were condemned to travel those dirt roads, as gravel and paving of all roads around the country would not come about until after World War I.

For hours the horses continued to battle a steady rain and the muddy roads. By midafternoon Bunk and Theo agreed that the horses needed a rest. They pulled into the driveway of the next farmhouse, hoping to find shelter until the drizzling rain stopped. The two men observed a young man plowing in a nearby field, whom they approached about permission to rest their horses for a while. The young man explained that he lived at the next place over, but he thought it would be just fine since the owner was away. He also told them they were welcome to spend the night in the vacant house. Being both cold and wet, they took him up on his offer. After putting the horses in the barn, Bunk and Theo carried their suitcases into the house, which was nicely furnished and very comfortable. The young man had also invited them to his home that evening. After eating their supper at the vacant house, Bunk and Theo joined the young man at his home for an evening of conversation. Returning to the empty house at ten o'clock, they set a nice corncob fire and went to bed pleased to be dry and out of the relentless rain.

Because of poor road conditions and frequent rain, it took them two days to reach Council Grove. Situated in the middle of the Kansas Flint Hills, the town got its name from a meeting of emissaries of the US government and the leaders of the Osage tribe in 1825. They met under a large oak tree to negotiate safe passage of white settlers traveling west along the Santa Fe Trail. The agreed-upon toll for that privilege was $800. When Bunk and Theo arrived in Council Grove on the fifth day of their journey, they agreed that they had had enough of the cold weather. During the day they wore several layers of clothes, but those cold nights without warm bedding had to end. At the general store in Council Grove they purchased a gasoline stove to warm the tent during the cold nights. Before they left the store, they purchased and mailed some postcards to their families back home.

Bunk remembered the Flint Hills of Kansas from when he traveled by train to Oklahoma a decade earlier. The area was nearly ten thousand square miles, with a rocky soil that would not permit farming but easily supported native grasses that fed tens of thousands of head of cattle. The rolling hills showcased changes in elevation up to hundreds of feet. As they drove the wagon across a high point in the Flint Hills, Bunk was amazed at being able to see for fifteen miles. That kind of terrain meant the roads were steep and the drive would exhaust the horses. It would take them two days to cross the Flint Hills. Bunk noted that without trees, the unrelenting wind made both travel and setting up tents difficult.

On the western edge of the Flint Hills, as Bunk and Theo approached the town of Marion, the rocky landscape transitioned to farmland, with lush fields of wheat and alfalfa. They had planned to set up camp for the evening before they reached Marion, but they could not find an area big enough to erect their tent because the crops were planted "right up to the wheel tracks." In Marion they rented rooms and put the horses in a livery stable. The next day they would leave the Santa Fe Trail and head south toward the Gypsum Hills on the southern border of the state.

For five days Bunk and Theo drove the team through the towns of Newton, Hutchinson, and Kingman before reaching Harper. A typical Kansas farming community located in Harper County on the Oklahoma border, Harper was dominated by a grain elevator and a railroad track. Surrounding Harper for as far as one could see were fields of new wheat. Harper County was first organized in 1873 under a cloud of fraud caused by three men described by A. T. Andreas in his *History of the State of Kansas* as "a groceryman and the other two soldiers of fortune — in plain English, men who lived by their wits." The men who conjured up the scam never actually resided in Harper County, but they understood how counties were organized and financed in those early days. They created the bogus town of Bluff City to be the county seat, a name possibly derived from their hope that no one would call their bluff as to the authenticity of the town. Next, they forged census documents purporting that hundreds of citizens lived in the county, listing names taken from a Cincinnati, Ohio, city directory. The last step in the fraud was to fake an election, which allowed the three men to issue bonds for public improvements, such as a county courthouse. Finally, the bonds were sold in faraway cities like St. Louis. When the dust settled, the three men absconded with $40,000 and Harper County would not be legally reorganized until 1878.

When Bunk and Theo arrived in Harper, they checked into the Patterson Hotel. After putting up the horses at the livery, they went to supper and then decided to investigate the town. Bunk wrote, "Took in the town, which ain't much." They found "some old soldiers on the street with fifes and drums making a great noise [and] attracting more attention than I think Harper had seen in a long time." Nearby, according to Bunk, "there was a lodge organizer on the street talking his head off and I went up and took ahold of his coat to get a better look at his button to see if he was a Woodman and thought I wanted a hand out and ducked." Bunk was amused by the man's sudden departure. The man's badge indicated that he was a member of the Tyrian-Ashler-Acacian, an early subgroup of the Freemasons. Bunk had hoped that the organizer was a member of the Modern Woodmen of America, an organization to which he belonged. Normally Bunk was not a joiner, but the Woodmen were a national fraternal organization of good repute that provided insurance to its members. Bunk always spoke highly of the organization and sought them out whenever he was traveling.

When the two men returned to the hotel for a good night's rest, Bunk began to look forward to specimen collecting in earnest in two more days.

CHAPTER EIGHT

Gypsum Hills and the Cimarron River

Bunk and Theo drove the wagon into the town of Medicine Lodge on the eastern edge of the Gypsum Hills just after nine o'clock in the morning of May 15, fourteen days after leaving Lawrence. Today that same 250-mile journey from Lawrence takes slightly over four hours by car. For eleven of those days they experienced unseasonably cold and rainy weather. The countryside slowly dragged by at about the same pace as a man walking. To break the monotony, they occasionally stopped to shoot and collect birds along the way. But in spite of those infrequent diversions, the better part of each day was filled with tedium and each other's company as they slowly crept toward their first collecting ground.

Medicine Lodge was adjacent to the Medicine River, the waters of which the local Native Americans considered medicinal. During the late 1800s the town became associated with a notorious former resident, Carrie Nation, a stern and disruptive supporter of the Women's Christian Temperance Union. Standing six feet tall and weighing 180 pounds, and best known for using a hatchet to break up saloons, destroying both their furnishings and alcoholic beverages, she presented a formidable figure. By her own admission, she was a "bulldog running along at the feet of Jesus, barking at what He doesn't like." Despite Kansas being the first state in the Union to enact Prohibition in 1881, enforcement was spotty. Otherwise, it is unclear why Carrie Nation's first attack on a saloon occurred in the nearby town of Kiowa in June of 1899, nearly two decades after Kansas outlawed liquor. Apparently, Barber County and the town of Kiowa, located in the very heart of cattle country, catered more to cowboys in the need of a drink than to the voters in the rest of the state.

At the general store in Medicine Lodge, Bunk and Theo bought a whip to replace the one they had lost on the road and a bag of chop to feed the horses. Afterward they filled their water can to the brim and headed toward the Gypsum Hills to find a suitable place to camp and hunt. Gyp Hills, as the area was often referred to, is semiarid, covers more than two counties, and is unlike anywhere else in the state. The bright red color of the soil and rocks reminded Bunk of the landscape near the Great Salt Plains in Oklahoma. Time and water have eroded the landscape, leaving hundreds of flat mesas, sometimes with a five-hundred-foot elevation drop to the valley floor. The red walls of the cliffs are accented with horizontal ribbons of white gypsum, accounting for the area's name. The canyons contain small streambeds, which occasionally carry water, enough to support elm, cottonwood, and cedar trees.

After crossing the Medicine River, the wagon immediately began to climb in elevation. At that time there were no real roads in the colorful canyons, only vague trails, so the horses labored to pull the loaded wagon. Bunk wrote in his field notes, "We stopped to camp at 1,700 feet according our conture [*sic*] maps." Without really noticing the climb, they had risen nine hundred feet in elevation since they left Lawrence.

Five hours later Bunk and Theo found their spot to camp, they put the exhausted horses out to graze, set up their tent, and began collecting specimens in earnest. Shotguns were the weapon of choice for hunting small birds, although one might think they would blow the petite specimens to smithereens. But even the smallest bird remains intact when the collector inserts a second barrel, called an "aux," short for auxiliary barrel, inside the primary barrel of the shotgun. Combined with light gunpowder loads and extremely fine shot, the aux assures that even the smallest birds can be hit at a low velocity and remain intact for the collection.

By late afternoon Bunk and Theo returned to their camp, each with a full bag of birds. They recorded their specimens in the field notes, then field dressed each bird by cutting open the belly with a knife and removing skin and feathers from the flesh, bones, and he internal organs. Next, they rubbed white powdered arsenic by hand on the inside of the bird skin to preserve the specimen from insects and bacteria. Last, they wrote the name of each bird and an identification number that matched a corresponding number in the field notes on a tag, attached it to the specimen, and boxed it up in a small wooden container to await shipment back to KU.

Bunk recorded in his field notes that first night his thoughts about the beauty of the Gypsum Hills. He described his setting by saying, "Our camp is almost surrounded by high red cliffs broken in to [*sic*] a hundred ravines and make the most picturesque camping spot I ever camped in[.] [There] is a watering hole close by fore [*sic*] horse watering and washing." Bunk also noted that both he and Theo went to bed with headaches. His notes offered no explanation; however, headaches are a well-known symptom and common side effect of arsenic poisoning. Considering the recent increase in their bird skinning and handling of arsenic, they most likely experienced an adverse reaction. In addition to headaches, excessive exposure to arsenic frequently causes skin sores—and in fact, several months later, Bunk lamented, "My face is broke out with arsenic so I can hardly stand it any longer[.] It has bothered me for three weeks and burns like fire all the time." Besides arsenic, the wagon also carried other toxic chemicals commonly used to preserve and/or euthanize animals: formaldehyde, denatured alcohol, and cyanide, to name a few. In those days, collectors and scientists did not fully appreciate the dangers of long-term and continued exposure to toxic chemicals. To them, the life of a collector involved tolerating a certain amount of pain and discomfort.

A few days later, Bunk and Theo left their camp and drove the wagon back to Medicine Lodge to pick up mail and supplies. Later that evening Bunk wrote, "It took just one hour and a half to make the same distance with the empty wagon . . . that it took five hours . . . with a load." When Bunk walked into the Medicine Lodge post office, he asked for any mail addressed to him. Without a permanent address in Medicine Lodge, he was looking for mail sent to him "in care of general delivery," meaning that a local post office held the mail until the addressee personally appeared and asked for it. Unlike today, when instantaneous communication requires nothing more than a push of a button, 1911 mail moved no faster than the speed of the trains that carried it from town to town. Whenever Bunk and Theo decided it was time to move on to another hunting ground, they always stopped by the post office on the way out of town to leave "forwarding instructions," asking the local post office to forward any future mail addressed to them to the next town on their route. Although slow by today's standards, the postal service proved more than adequate. Through that simple system, any mail sent from family or the university to any post office along their route simply followed the two men across the state.

On that day, Bunk received a letter from McClung asking that the contents of the stomachs of all the mammals collected be preserved so they could see what the animals were eating. The request amused Bunk, because they had not seen many mammals up to that point. "I don't think it will take a very large jar to put them in as we have not seen any signs of any in these barren hills." McClung's letter never mentioned whether the funding from the Kansas Biological Survey had been approved or not. Bunk considered that good news, because if the request had been denied he would have been instructed to return to Lawrence immediately.

Bunk also received a letter from Clara telling him about Audrey's first birthday. As the two men rode back to camp, Bunk thought about how he regretted missing his daughter's birthday. But like many men in that day, Bunk never spent much time at home dealing with family matters. He devoted most of his time to his job at the university while Clara managed the home front. For years Bunk generally worked long hours, usually six days a week, so he knew Clara was accustomed to his absence. But this trip was considerably longer and would be especially difficult for his wife. Although he missed his wife and two daughters, as usual he felt his work was a priority.

When they arrived back at their campsite, Bunk gathered up his towel and soap and headed toward the nearby stream for a bath. He couldn't remember the last time he had taken one, but he knew he was overdue. Checking the temperature of the water nearly caused him to change his mind about the bath. Bunk soaped and rinsed himself hurriedly, but as he dried off and the warmth returned, he was glad to be rid of so many layers of dust. Whenever Bunk and Theo went camping, clean clothes were infrequent and bathing a rarity. To feel refreshed, the best they could usually hope for was a sponge bath while standing in the tent or going to town for a shave from a barber. Consequently, most days both men looked pretty scruffy.

For five days, Bunk and Theo walked the red hills around their Gyp Hill campsite in search of birds. On average they collected twenty-five birds a day. Bunk had expected to do better, but the strong winds caused the birds to hide in the protection of the trees and bushes. On those rare occasions when the birds left their cover, the wind made it difficult if not impossible to shoot with any accuracy. At night the high winds continued, pulling tent pegs out of the ground, tearing the tent, and making the fly flap continuously. Furthermore, the wind quickly dried up what little water there was in the creek, so driving to town for water for themselves and the horses wasted valuable time away from hunting.

After a few days, the weather changed, and cool winds brought sporadic rains. Bunk appreciated how a good rain would fill up the dry creeks and replenish their water supply. On the other hand, rain hampered the hunting efforts during the day and a large downpour might flood their campsite. At one point he recorded that concern, writing, "We might have to move in a hurry as we are camped in a little flat in a bend of a dry creek that drains a good deal of this hill country."

When the bird hunting began to play out, Bunk decided to move their camp further west to a little town named Aetna, Kansas. Even though it was only ten miles away, it took two days to reach the town because of the hilly terrain and sandy soil. Today the only trace of Aetna is one small and dilapidated old building sitting on a gravel road at a T intersection on a high plateau surrounded by sage brush and thistle. Bunk's description of the town was: "Just a hotel, 2 stories [tall], a store, a barn and no houses. And what a store it was; plenty of room and shelving, but empty as a barn. The storekeeper acted rather strange and we did not understand him. Did not have bacon, canned goods, sugar, in fact a few crackers and eggs was all he had to eat."

Bunk asked the clerk if they needed permission to hunt around Aetna. The clerk replied that the rancher who owned most of the property lived about two miles south of town, across the Salt Fork River. They drove the wagon to his house and found his children at home, who said their parents were out fishing for supper. After driving around for an hour, they located the rancher and his wife fishing on a small creek. At first the rancher hesitated to give them permission, but eventually they convinced him to let them hunt because it was for scientific purposes. With permission secured, Bunk and Theo drove to an area the rancher recommended called Water Canyon, where they were pleasantly surprised to find water, grass, and trees. They set up camp on Cave Creek just downstream from Water Canyon. For the first time on the trip, Bunk saw signs of mammals near their camp. That evening he set out traps and collected several mice the following morning. For the remaining few days of their stay near Cave Creek, the mammal collecting was successful.

By the end of May it was time to move on to their next collecting field, the Cimarron River. Their actual departure would depend on the weather. If it continued to rain, the soil would be too wet for the horses to pull the wagon. Hopefully the wind and sun would continue to dry their path out of the hills. Some years in Kansas the transition from a cold spring to boiling summer occurs instantaneously, and 1911 was no exception. Since the first of May, it seemed to Bunk and Theo that they had constantly shivered with the cold rainy weather. Now in late May, as they were preparing to leave the Gypsum Hills, the rain stopped abruptly and, without missing a beat, the summer heat came on with a vengeance. For the next two days the sun and blowing wind dried out the sandy soil around Cave Creek. This meant they could exit the Gypsum Hills; unfortunately, the heat would also make their travel considerably more unpleasant.

On the morning of May 30, after waking at six o'clock, Bunk and Rocklund ate breakfast, packed the wagon, and readied the horses. Although the ground had dried enough, the sandy footing proved to be a problem. Sweat already poured off both men by eight o'clock, when they mounted the wagon and signaled to the horses. Billy and Baldy immediately struggled to pull the fully loaded wagon through the soft sand of the deep ravines. After a quarter mile they came to what Bunk described as a "mean little gully." They looked over the terrain and concluded that they needed to improve the "going out side" of the ditch with their spades. For thirty minutes they toiled in the hot sun, reshaping the gully. When Bunk flipped the reins, Billy and Baldy cautiously pulled the wagon into the dry creek bed. As the horses strained to climb out the other side of the gully, one of the back wheels slipped sideways. With a loud *thwack*, the wagon tongue snapped completely into two pieces.

Bunk and Rocklund realized that, being miles from any town, they were on their own to fix the wagon. Worried, but not panicked, they came up with a temporary solution of splicing some slender branches from a nearby elm tree to the tongue using some quarter-inch rope. The repair was sufficient to allow them to drive the wagon out of the canyon and over a divide to another creek. Stopping to eat dinner, Bunk took a photograph of the spliced wagon tongue, noting, "I was proud of my job."

Fifteen minutes after they resumed their journey, they came to another ravine. As the horses struggled to pull the wagon out of that creek bed, once again the sandy soil caused the front wheel to turn into the bank of the gully. Between the weight of the wagon and the pull of horses, the newly repaired wagon tongue broke again, but this time in a different spot. As the horses trotted forward with the wagon tongue, harnesses, and reins in tow, Bunk and Theo sat there dejected. In Bunk's own words, "then we were in it."

They were miles from the next town and in a place where they would never see another person for weeks. "We took the horses off or rather took their harness off as they were all ready [*sic*] off, unloaded our wagon and stood and looked at our outfit for several minutes." Thinking there was certainly a ranch on ahead of them, Bunk started walking to get some help. After about a mile, he found a ranch. Near the windmill stood a woman and two young girls loading barrels of water onto a wagon. Bunk told the

Bunk beside wagon with broken tongue, 1911, west of Gypsum Hills (University of Kansas Natural History Museum records, University Archives, RG 33/0 Photographs, Kenneth Spencer Research Library, University of Kansas Libraries)

woman of his problem and asked if she had extra wagon parts. She led him over to the barn, where he saw some parts that he thought he could use to make the repair. Before she could give Bunk permission, though, he needed to talk with her husband, who was out mending fences. The woman gave him directions and informed him that she and the girls would be gone when he returned — they were taking the water barrels to "drown out prairie dogs, as the girls had been after her all day about it."

After a short walk, Bunk found the rancher and explained his dilemma. He was willing to rent or buy a tongue or pay him to drive him to the closest town. The rancher explained that he was worried that his cattle would get into his corn field if he did not mend the fence that day. In the meantime, the rancher welcomed Bunk to work on the repair back at his barn. Bunk thanked him and walked back to the wagon. He and Rocklund removed the tongue and carried it to the rancher's place. Since the wagon tongue at the ranch did not match theirs, they had to combine and adapt assorted wagon hitch parts. The repair involved drilling holes in metal strapping and attaching it to the broken parts with screws. After they returned to camp after dark, they ate a late supper. It was half past eleven when Bunk finished writing his notes for the day.

After a night of fitful sleep, Bunk and Rocklund awoke at 4:30 a.m. and

Repaired wagon tongue (University of Kansas Natural History Museum records, University Archives, RG 33/0 Photographs, Kenneth Spencer Research Library, University of Kansas Libraries)

were at the ranch house before the rancher woke. Once Bunk and Theo finished repairing the wagon tongue, they settled up with the rancher and carried the new tongue back to their wagon. Since they knew the day was going to be a scorcher, they topped off their water can with cistern water from the ranch. By nine o'clock that morning they were on the road again.

The temperature and humidity began to rise as they drove the wagon toward the town of Coldwater, Kansas. Bunk commented, "It was intensely hot today and we drank an extra lot of water at noon." After dinner both men became nauseated. Bunk surmised that the water from the cistern at the ranch was tainted. Bunk wrote, "I was the worse off and could hardly stand it to ride. In fact I got off the wagon several times and walked in the boiling sun to try and settle my stomach." In spite of their discomfort, they had no other choice but to continue on.

Late afternoon they arrived in Coldwater and established a campsite in a pasture south of town. Bunk "got on a cot as soon as we got the tent up." After an hour or so, he decided to walk to town hoping to find something to settle his stomach. "On the way to town I threw up my breakfast and dinner." In town he bought a ginger ale, got a shave, and mailed some letters. As he started back to camp, he was feeling so poorly that he mistakenly walked north instead of south. Figuring out his mistake, he walked back to town and "started over." When he arrived back at camp at nine thirty that night, Theo was relieved to see him and told Bunk he would have gone looking for him had he not felt so sick himself.

Still feeling weak and puny the next morning, Bunk and Theo thought they should probably eat something to alleviate the nausea. They drove the wagon into town for breakfast. After eating at a café, they bought some wooden boxes for shipping bird and mammal specimens back to Lawrence on the train. Bunk reported with a note of optimism, "We took it rather easy today on account of feeling bad, but got things in shape to make an early start tomorrow for Protection and the Cimarron River, where we expect to spend ten days."

Protection, Kansas, was a small town about eight miles west of Coldwater. The town's claim to fame would come several decades later, when the new polio vaccine was first introduced. The town's name seemed appropriate for those first inoculations, and as a publicity strategy, the National Polio Foundation selected it as the site where it would vaccinate every resident under the age of forty, resulting in the first city population to be protected from the disease. When Bunk and Theo arrived, they found a baseball game in progress. They stopped to watch for a while and learned that two other games were scheduled for the afternoon. Bunk was amazed that so many people were there to play baseball even though Protection had only a few hundred residents. He reported in his notes, "There were probably fifty out-of-town automobiles in town for the games." When they left Protection, Bunk decided to take it slow, as Baldy's shoulder was causing the horse to limp, which he blamed on the hard pulling the horses experienced coming out of the Gypsum Hills.

When they turned south at Ashland, the hot sun was high overhead. With a sandy roadbed and "a bad south wind that was full of fine sand," both man and horse labored. Just two miles short of the river, the road surface suddenly became more drivable. Now the sand in the road had been mixed with clay and then graded. After a mile, they discovered the reason for the improvement. In the middle of the road sat a small house and toll gate where a young boy collected twenty-five cents for a round trip. Appreciating the better road condition and not having another option, they paid the toll and drove on to the river.

Bunk and Theo arrived at the hunting grounds on the Cimarron River on June 3. When the wagon pulled up to the north side of the Cimarron River, Bunk was stunned. Ten years earlier he had collected on this very same river, but in Oklahoma. There it flowed through a lush and beautiful piece of land. He could hardly believe it was the same river. Here the shallow water of the Cimarron River slowly meandered through muddy, gravelly sandbars in the riverbed. A treeless stretch of scrub brush surrounded several tiny creeks that fed a nearby boggy wetland. A one-lane wooden bridge spanned the river, connecting what Bunk described as an area where "there are no trees in sight and very few plumb [*sic*] bushes. It is the most desolate looking piece of country with only sand hills to break the view to the next county."

Despite its appearance, this isolated hunting ground offered ample wildlife. Every morning Bunk would harvest numerous sand rats and mice from the traps he set out the previous night. After breakfast the two men would shoot thirty to forty birds, which they would skin after dinner. Bunk and Theo observed mink and muskrats as well as sighting coyote tracks. The shallow river was home to numerous fish and turtles, and hundreds of cliff swallows that used the river mud to make their nests under the bridge.

Even though their collecting was successful, the area was hardly idyllic. In addition to oppressive heat and hard dry winds, flying, biting insects were everywhere. On the second day at the river, Bunk went looking for a better campsite and found one he thought would be an improvement. When he was suddenly attacked by thousands of mosquitoes and large biting flies, he decided that the initial campsite was satisfactory after all.

On the evening of the second day in camp, the boy from the toll house and his father, the toll keeper, stopped by to visit and say hello. According to Bunk, the man had "been a cattleman and westerner from childhood and is a gentleman as all such are that I have ever met." Their conversation with the man included the topics of how the prairie chickens were breeding in the area, how plentiful the quail were, and how the migratory ducks destroyed crops if they arrived before the grains are harvested.

That part of the Cimarron River was also cattle country. When the pasturing cattle searched for water, they often visited the campsite. In fact, Bunk complained in his notes that "they bother us some by drifting about our tent and wagon and occasionally getting mixed up in the ropes[.] You can not [*sic*] scare them nor run them far. They stop running when we do and come right back."

One day Bunk decided to collect some cliff swallow eggs. He climbed down the river bank and waded into the shallow water under the bridge. To his surprise, many of the mud nests were "broken down." He surmised that the extensive nest damage resulted from the heavy lumber truck traffic they had seen on the road. He also concluded that the presence of commercial trucks explained why someone would think to charge a toll on what seemed to be a road to nowhere. Regardless, Bunk still thought it all very strange because there were no houses or buildings that he could see for miles.

On their last day along the Cimarron River, the toll road keeper and his family came by to invite Bunk and Rocklund to go fishing over on Bullard Creek, about six miles west of their camp. Bunk had no interest in the fish, but he thought they might get some turtles for specimens. When he arrived at the small creek, Bunk questioned the location, because "it did not look like it had anything bigger than a minnow in it." Much to his surprise, several sizable carp, channel catfish, and mudcats occupied the stream, and the toll keeper and his family caught enough to provide a "nice platter of fried fish" that evening for everyone. Instead of fishing, Bunk and Theo caught forty turtles. Just like back at their Cimarron River campsite, swarms of mosquitoes hounded everyone. "The women and children and some of the men suffered as the bites swelled up on them like hives. We have gotten use [*sic*] to them so we can get bit about five thousand times without minding it much."

On the morning of June 12, 1911, Bunk and Theo broke camp at the Cimarron River and drove back to Ashland. They intended to stay longer on the Cimarron River, but the heat and dust as well as the wildlife began to play out. They stopped at the blacksmith shop and, in Bunk's words, "got our wagon tongue fixed better as we were afraid of it the way we had it patched." When they left town, they traveled northwest along Beaver Creek. The road elevation climbed up steadily and by the end of the day their camp was four hundred feet higher than Ashland. From the lofty vantage point, they could see for miles back over the hills they had left behind. Bunk wrote that evening, "We are camped tonight on a high rolling piece of country where we can look back over as pretty piece of cattle country as we have seen."

As they rode on to Wallace, Kansas, Bunk's field notes transitioned from stories about collecting to his experiences on the road. He recollected how the wheat fields gave way to a treeless countryside with only sparse grass and scrub brush. They bought hay for the horses after learning that there would not be enough grass to sustain the horses as they continued west. When Bunk passed through Ulysses, Kansas, they saw fancy sod houses with second stories, gabled roofs, and wallpapered interiors. They found the land so flat that they got lost on one occasion because their contour maps failed to provide enough landmarks. When the wind picked up that evening, the tumbleweeds were "chasing each other towards the horizon in rapid succession," causing Bunk to think the weather was shifting.

Bunk wrote that they were going to "new" Ulysses, because the original town had been located two miles to the east. The town was moved, literally, when the citizens were swindled by some carpetbaggers who sold bonds for establishing the county seat and absconded with the proceeds. For decades the townspeople tried to honor the debt; however, a few years before Bunk and Theo arrived, the population was so small that they could not afford the payments any longer. To solve their dilemma, the residents moved all the town buildings to a new location by putting them on wagons and skids and pulling them with horses. Even the two-story hotel was cut into sections and moved to the new townsite.

Bunk wrote about their encounter with a powerful, fast-moving storm along the road north of Johnson City, Kansas. They noticed it "just as a big cloud of dust was almost on us. Before we had time to put anything away, it sent things flying and filled our water pail and cups of lemonade and jar of jam with dirt and trash." Once they had everything from their dinner in the wagon and the horses were hitched, they forced the wagon north into the face of the storm. Suddenly a cold hard rain hit them as quickly as the dust storm hit them earlier. Luckily the two soggy travelers only had to drive about a mile before they found a dilapidated and deserted cement house once occupied by early homesteaders. Realizing that the rain would likely continue, they took the harnesses off Billy and Baldy and put them out to graze. Bunk wrote that they "brought in our gas stove, a table and cots and got supper. We tacked a piece of canvas up to the window that was not boarded up and left the gas stove burn to take the chill off. We are as comfortable as if at home."

In Syracuse, Kansas, they ran into a man on the street whom Bunk described as the "town liar." The man asked why they were in town and upon hearing their explanation he monopolized the conversation. As Bunk explained, the man "came around and told us of all the big game he had killed, fine shots he had made and places he had been in [the] early days until we could not stand it any longer." Plain-talking Bunk had little tolerance for braggarts, so they abruptly walked away. After supper, the town liar found them again and bored them until ten o'clock that night, when they again excused themselves and walked back to the livery stable. They unloaded their cots from the wagon and went to bed next to the livery stable.

That night Bunk recollected the time they had spent since they pointed the wagon out of Lawrence on the first of May. They had collected hundreds of birds and mammals and traveled hundreds of miles. Both men and their horses were tired and dirty but had survived the extreme cold and heat and endless camping. In spite of his fatigue, Bunk appreciated the diversity of the Kansas landscape: the fertile fields of wheat and alfalfa, the undulating Flint Hills, the red rock country of the Gypsum Hills, and the vast high plains along the western border of the state. Unfortunately, he also recalled the long hours and excruciating heat when he and Theo repaired the broken wagon in Gypsum Hills, and the dust storm that turned into a torrential rain on the drive to Syracuse. As he dozed off, he was ready to move on to Wallace.

It took them four days to reach the small town of Wallace, where they planned to meet Martin and the fossil team. Bunk wrote in his field notes that they wanted to say they'd "been in Colorado," so when they left Syracuse they pointed the wagon west. Six miles from the state line, they drove through Holly, Colorado, before heading northeast to Tribune, Kansas.

CHAPTER NINE

Wallace County

When Bunk and Theo arrived in the town of Wallace on June 23, 1911, they were looking forward to handing over the horses and wagon to Handel Martin and the fossil party from KU. As the two men bumped along in the wagon, they talked about how they were eagerly anticipating taking the train back to Lawrence for well-earned rest and relaxation. Despite their fatigue and exhaustion, they were filled with a sense of pride for the significant number of specimens they had collected, especially birds.

The town of Wallace was located a few miles from the eastern edge of Wallace County. Being on the treeless, high plains of western Kansas, by midsummer the landscape radiates like a furnace. The town consisted of a several stores and a few dozen houses that straddled the Union Pacific Railroad track that connected it to the rest of the world. Theo reined the exhausted horses to a stop in front of the post office, where Bunk hoped to find a letter waiting for them from Martin naming the person to contact for directions to the fossil field. Once there they would set up camp and wait for Martin and his crew. Much to his disappointment, the postmaster informed Bunk that there was no mail for him.

The absence of a letter from Martin meant Bunk had to improvise. Figuring that there weren't too many fossil collectors in a town that size, he asked the postmaster, who suggested he look up a local man by the name of George Allaman. Following his directions, Bunk and Theo drove the horses and their wagon to Allaman's place, a mile and a half straight south of Wallace. They found the setting quite picturesque, a ranch house situated in a grove of trees along the slow-moving, shallow Smoky Hill River. Bunk explained their business and Allaman agreed to let them camp on his ranch, but told them they were not allowed to shoot birds. That restriction both surprised and disappointed Bunk, but he reluctantly agreed, since he expected Martin to arrive soon.

Allaman suggested that they follow the Smoky Hill River about a half mile east of the house. They could camp along the river, where some tall cottonwood trees would give them shade and there was plenty of grass for the horses. In fact, the grass was so long that Bunk and Theo had to cut some down at the tent site before they could move their cots in. Although Bunk admitted that it was a comfortable campsite, he suspected that he would miss the opportunity to collect birds. In fact, several days later, he lamented, "Trapping and looking for snakes is tame compared with hunting the feathered tribe."

After setting up camp that first afternoon, Bunk and Theo walked back to Allaman's home that evening to join his family for supper. The conversation was lively and wide-ranging, with Allaman at the center of it all. As they visited, Bunk soon began to understand that George H. Allaman was the elder statesman of that part of the state — more than just an average farmer and rancher. In rapid succession, Allaman told his stories with a glint in his eye. He confidently talked about any number of subjects, including fossils, western Kansas agriculture, and the "old days" when Wallace County was first settled. This was particularly true when Allaman had a new visitor, especially one like Bunk, who was from the university and appreciated the significance of his stories. When Allaman mentioned that he had worked with Martin, Wendell Williston, and Charles H. Sternberg, Bunk realized that his new acquaintance was a great source for information about local fossil hunting. In addition, Allaman was well versed in many subjects relating to early days in western Kansas. He was one of the first settlers of Wallace County, having arrived at the time of the building of the Union Pacific Railroad. Initially he made his living scouting for the army in the Indian Territory around Fort Wallace, a few miles from the town of Wallace. Later he hunted buffalo and antelope, selling the meat as food for the railroads' construction crews and the soldiers at the fort.

Allaman's old stories enthralled Bunk. One such story took place in 1876, when Allaman traveled to Kit Carson, a small town in eastern Colorado, to help a man sell his ranch. While he was there he met and dined with General George Custer just a month before the ill-fated Battle of the Little Bighorn. Bunk shared with the rancher that the lone survivor of that battle now resided at the KU Museum of Natural History. Allaman talked about how he became interested in ranching, got involved in local politics, and started a bank. He also explained how before he arrived on the high plains of western Kansas, no one had successfully grown crops there because of the lack of rainfall in that part of the state. Allaman explained that he experimented by irrigating crops with well water. Eventually, he successfully grew a number of crops, including alfalfa, fruit trees, melons, onions, celery, and cabbage.

With all the discussion about agriculture and growing crops, Bunk realized why they had not been allowed to shoot birds. Allaman understood that birds protected the crops by eating the harmful insects that caused damage. (Ironically, many weeks later, when Bunk and Theo were departing Wallace, Allaman asked them to shoot some birds that were eating the fruit from his trees.) Allaman's interest in science and his wealth of stories about the early days captivated Bunk so much that he went by to visit Allaman almost every day during their stay in Wallace.

For the next few days Bunk and Theo hunted everything but birds. When the game seemed to be played out, they went to look for a new spot to camp. The next ranch down the river to the east was owned by a Kansas City banker named Pinnell. For permission to camp there, they needed to see the foreman, Mr. King, and they spent half a day searching for him. King gave them permission to camp, turn the horses in to the pastures, and hunt for anything, including the birds, which was what Bunk wanted to hear.

On the fifth day after their arrival, June 28, Bunk and Theo drove the wagon to Wallace to pick up mail. This time the postmaster had a letter for them from Martin saying that the fossil men would not arrive until Tuesday, July 10. Feeling the effects of collecting for nearly two months, Bunk became resigned to the situation when he wrote, "I guess we can stand it as we have got used to the weather and camp life and our own cooking which is mostly warming up canned goods."

The following day, Bunk walked over to Allaman's place to let him know of Martin's delay. As usual, Allaman was very chatty and began to show Bunk some fossils and stone carvings. Bunk asked about good places to collect fossils. Allaman suggested a worthwhile place to hunt on the Pinnell ranch and gave Bunk directions to a canyon on the south side of the river about a mile to the east.

After Bunk left Allaman's place, he walked back toward the Pinnell ranch, continuing to look for birds. He could not stop thinking about what Allaman had said about the fossils. By his own admission, Bunk was more interested in birds and mammals, but his time spent around paleontologists had given him valuable exposure to the world of fossils. He remembered Williston telling him that great fossil hunters, like Sternberg, have an uncanny ability to recognize unusual or irregular horizons in geological formations that suggest the presence of fossils. Martin had shown him how to look at

Campsite on the banks of the Smoky Hill River south of Wallace, Kansas, 1911 (University of Kansas Natural History Museum records, University Archives, RG 33/0 Photographs, Kenneth Spencer Research Library, University of Kansas Libraries)

a piece of rocky rubble and distinguish rock from fossil by the difference in color or texture. As it turned out, this time waiting for Martin gave him a chance to work on his fossil hunting skills.

Nearing their campsite, Bunk found the canyon described by Allaman. According to Bunk, that was the moment when "fossil hunting got the best of me." For several minutes he poked the ground around a little knoll in the canyon. Suddenly he noticed a few fossil fragments, which caused him to hasten and widen his search. With more poking around in the dry, rocky ground, more fossil pieces appeared. Bunk thought those fossils and their arrangement on the ground indicated that he might be looking at a prehistoric marine reptile, a plesiosaur.

When Bunk first arrived at KU, he observed that many of the fossils in the collection were aquatic and had come from western Kansas, which seemed rather odd, considering that they were found thousands of miles from any ocean. When Bunk asked Williston about it, the professor explained that sixty million years ago Kansas and much of the Midwest had been covered by the Western Interior Seaway, a shallow saltwater body that stretched from today's Gulf of Mexico to the Arctic Circle. When animals living in the ocean died, they sank to the bottom and became part of the salty ooze and muck that would turn into limestone. Eventually the water of the inland

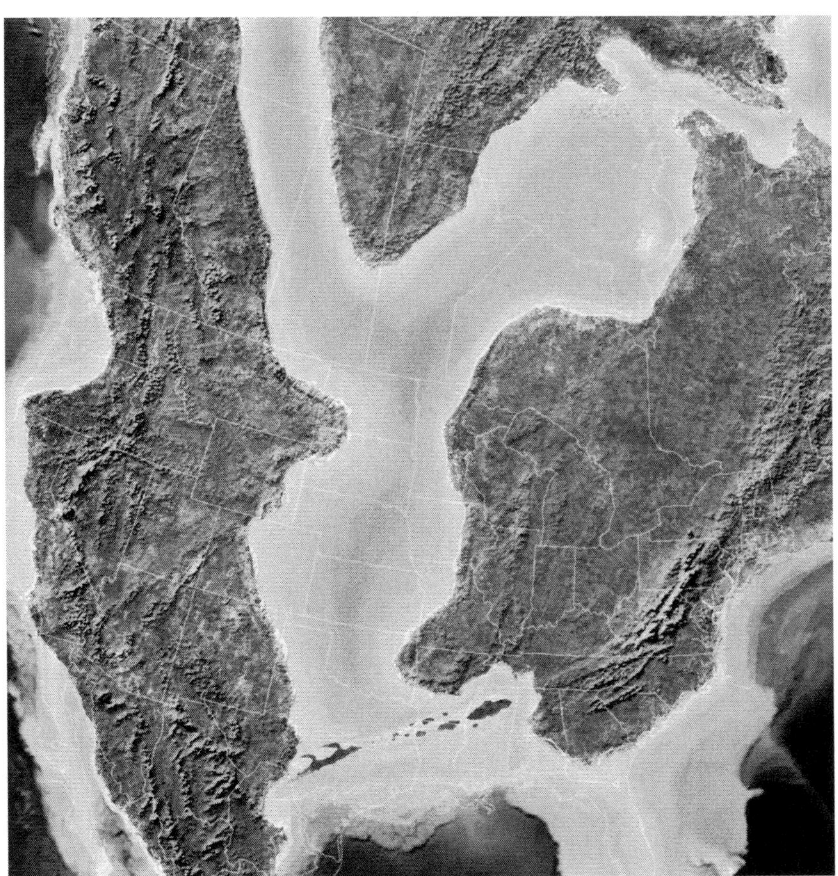

Western Interior Seaway (permission and courtesy of University of Kansas Natural History Museum)

sea drained when the Earth's tectonic plates collided. That geological movement, which occurred over millions of years, caused much of what we know as the western part of the United States to rise up, creating the Rocky Mountains and the current Pacific shoreline. As part of that geologic shift, what had once been the ocean floor was now forced to the surface. That resulted in rocky outcroppings that contained a wealth of marine fossils.

The plesiosaurs' taxonomic family consists of half a dozen distinct genera (the plural of *genus*) and nearly a hundred different species. In general, plesiosaurs had a rounded, torpedo-shaped body with four flippers, some with long necks and others with short ones. As cold-blooded reptiles, they were attracted to the warmer waters of the shallow inland sea. Plesiosaurs ranged in size from five to fifty feet in length. Compared to the rest of the creature, their heads were small, but their long jaws contained rows of razor-sharp teeth that intermeshed. They were effective predators, especially those with long necks that enabled them to get in the middle of a school of marine life and dine before their body mass would arrive and scare away the prey.

Despite the excitement of finding the marine fossil, Bunk did not have the proper tools with him for digging, so he reluctantly went back to camp. Typically Bunk would rather hunt birds than anything else, but the idea of a large fossil had him keyed up. After breakfast the following morning, Bunk showed Theo the fossil site and the two men began looking for more bone fragments. While Bunk dug in one spot, Theo moved down the line to search the wall of the rock canyon. Theo soon returned with some pieces of what Bunk surmised were part of a jawbone. This was a choice fragment and indicated to Bunk that this site had not previously been worked, because no collector worth his salt would ever have left "the end of the jaw."

The soil where the bones were uncovered was brittle and contained broken stone. Bunk deduced that this soil must have been a soft mud at the time the bones were covered, since all cavities were filled with it and the substance was now harder than the bone. Such conditions made it impossible to take the bones up in one piece, so they had to collect mostly broken fragments.

The next day was July 1, and Bunk and Theo agreed they needed a break from the strenuous digging and hot wind. They hitched the wagon and went about five miles to the southeast to visit two different canyons. When they could not find any fossils, they returned to camp. In the afternoon, after their midday meal, Bunk took a walk to look for turtles, finding fifteen. Theo spent the afternoon hunting birds and shot some burrowing owls and a ground squirrel. Later they stopped to see King at the Pinnell ranch. He offered them some daily newspapers, which were the first they had read since they left the comforts and daily routine of home two months earlier. That evening Bunk wrote, "We had almost forgot there was such a thing as news from the outside world."

Life on a specimen expedition has its periods of boredom, so Bunk and Theo appreciated how much Billy and Baldy entertained and amused them. Bunk wrote in his notes that when Theo was a little late in feeding Billy his chop, at least according to the horse, Billy "will come up to the wagon and stand around until Theo gets his box ready for him and if Theo tries to walk far with the box Billy will get in front of him and try to make him put the box down." Billy pretty much had the run of the Wallace camp because he was no threat to run off. Baldy, however, was another matter. To keep him in check they always had Baldy tethered to a picket line, which allowed him to graze but not run off.

On Sunday, July 2, they took the wagon back to the fossil site. They resumed digging and excavated more fragments. After several hours, they had dug the entire top off of the knoll and found a second specimen, which was much more fragile than the first.

Wanting to take a breather from digging, they carefully loaded all the fragments into the wagon and started back to camp. Because the fossils were so fragile, they drove slowly so the specimens would not break. Once back in camp, they loaded their fossils in a wooden box for shipping the specimens back to Lawrence. They packed up all of the plesiosaur fragments, as well as some coyote skulls and seashells, plus some dirty clothes to fill up the empty spaces in the box. The men discussed how they were quite proud of their fossils, but little did they know a bigger prize was in their future.

After dinner, the afternoon heat was fierce and the birds were scarce. Bunk and Theo decided to take their small pick and walk up into a little dry canyon where they had not been previously. Bunk found nudging and stabbing the ground with the pick to be much less strenuous than digging, and a nice break from the morning activity of building and packing the shipping crate. After about half an hour of poking around, Bunk found some exposed fossilized bones embedded in the side of the canyon. He immediately recognized that the anatomy of these bones was very different from the plesiosaur that they had been working on in the other canyon. He also realized that these fossilized bones were much larger and seemed to cover more ground, indicating that this prehistoric creature was much larger than their earlier find.

Even though they were excited about this new fossil, again they were not really equipped with the right tools for removing large amounts of rock and soil, so they decided to leave and come back later. They marked the spot with Bunk's handkerchief to make sure they could find it when they returned with the proper equipment. That evening Bunk made his daily entry in his journal. Although he was excited about the fossil, in his typically unemotional manner he called it simply "a very large saurian," a generic name that includes everything from present-day lizards and crocodiles to extinct species of dinosaurs (terrestrial), ichthyosaurs (aquatic), and pterosaurs (terrestrial and capable of gliding and flying). It would take more digging to identify the fossil.

The morning of July 3 brought calm breezes. Without the stiff winds of the previous few days, the bird hunting could resume. As they hunted through the canyons around their camp, Bunk continued thinking about the large fossil. By midafternoon they had dressed, measured, and tagged thirty-one birds. With that completed, they drove the wagon to Wallace to ship the boxes of specimens to Lawrence and pick up mail; a letter from Clara informed Bunk of Martin's departure from Lawrence.

When Bunk woke on Tuesday morning, he was looking forward to the Fourth of July picnic that afternoon. After breakfast, Theo puttered around the camp tending to the horses, cleaning guns, and reloading shotgun shell casings. Because it was too early to get ready for the picnic and Bunk could not wait any longer, he picked up his spade and walked to the location of the new fossil. After a short time digging, he found what he thought was a large continuous piece of bone. As more of the bone became visible, he recognized it as jawbone. He wanted to measure it; however, he had left his cloth measuring tape back at the camp, so he used the only thing he could think to calculate the length. In his notes he declared that the jaw was "six inches longer than my spade, handle and all." Given that the jaw was at least four feet long and contained three-to-four-inch teeth, Bunk knew this fossilized sea serpent was a significant find. In spite of his excitement, he was unable to offer a hint of exhilaration or elation in his field notes. In his typical understated manner, he wrote that he "came back to camp and cleaned up a little" for the picnic.

CHAPTER TEN

Excavating the Giant Sea Serpent

When the two weary collectors from the university arrived at the Fourth of July picnic, they found what Bunk described as "a good crowd of good natured people of the West." The Modern Woodmen of America had organized the picnic and held it half a mile east of Bunk and Theo's campsite. Bunk wanted to share his experience of finding the newly discovered fossil with George Allaman, so he eagerly sought him out among the crowd. As soon as they found him, Allaman invited the two men to join him and his wife for dinner. In his field notes, Bunk wrote, "We did and what a dinner; fried chicken, great platters full, cherry pie, raspberry pie, apple pie, cake, pickles, ice cream[,] coffee and all sorts of good things."

After dinner, Bunk finally had the opportunity to tell Allaman about the large fossil they had found. Bunk described the size of the jawbone and although only part of it was exposed in the rocky outcrop, he suggested that the jawbone reminded him of the mosasaur skull that he had worked on for Charles Sternberg five years earlier. Allaman firmly offered his opinion that "any saurian found in those hills, would be plesiosaurs and not mosasaurs." Although Bunk did not consider himself an expert in paleontology, he was fairly certain that his fossil was a mosasaur. Out of courtesy to his host, he decided to drop the subject.

As Bunk and Theo walked back to camp after the picnic, Bunk couldn't keep from thinking about the new fossil. He told Theo that he was hoping that it was a mosasaur because they were considered such an extremely ferocious predator from the same prehistoric era as the plesiosaurs. Mosasaurs were first discovered in the Netherlands in 1764 at Maastricht near the Meuse River. It wasn't until 1822 that they were named scientifically—*mosasaur*, which literally means "Meuse River lizard." Many early naturalists thought that mosasaurs were land creatures because of their resemblance to a monitor lizard. By the middle of the 1880s naturalists realized mosasaurs were sea creatures because what were originally thought to be legs were more likely flippers. By the end of the nineteenth century mosasaurs were being found in North America as well as Europe in marine deposits from the late Cretaceous when the Western Interior Seaway covered much of the continent and Europe existed as a chain of islands surrounded by warm shallow water.

Like plesiosaurs, mosasaurs evolved as an aquatic reptile from a terrestrial one. They were air breathers — they had to swim to the surface to fill their lungs — and gave birth to live young rather than lay eggs. But that is where the similarities ceased. A mosasaur generally resembled today's crocodile, only bulkier. They had a long tail, short neck, and a long jaw filled with sharp pointed teeth that crushed and killed their quarry. They dined primarily on fish, sharks, sea mollusks, or occasional sea birds floating on the surface. Because they lacked bladelike teeth, they were unable to rip flesh from their victims, so they swallowed them whole. Being warm-blooded required them to eat frequently for warmth and to support their high level of activity.

Allaman had only offered his admonition about the species because he had never known anyone to find a mosasaur in that part of the state, but that didn't make it so. In any event, Bunk would rely on the paleontologists at KU to identify his fossil. His job would be to get it home.

When Bunk and Theo finally reached camp, it was late in the afternoon. Much to their dismay, they discovered that the horses were gone. Frantically they searched the perimeter of the camp, where they found horse tracks that indicated that Billy and Baldy had gone north toward Wallace. It was getting dark, so Theo worked on supper while Bunk began searching to the north. When he came to a fence near the Union Pacific railroad tracks he found there a closed gate, so it was unlikely he would find the horses in the fenced area. When he looked to the east he could faintly see two horses running away from him. But it was getting too dark to see, so he decided to call off the search until the next morning.

Worried about the horses, Bunk had trouble sleeping that night. If the horses got hurt, their collecting expedition would be in a fine mess, and the longer the horses were away, that was collecting time they could not make up. He woke up early, at four thirty the next morning, to search for the horses. He walked back to Wallace and the gated pasture and began to walk the perimeter of the fence to see if it had an opening. Sure enough, he found where the fence was down. He saw a herd of horses almost a mile away, near a house. When he got closer, he saw the white face of Baldy.

As he slowly approached the horses, they began to move off, but not before Bunk grabbed the long rope from the picket line that Baldy had been dragging. Since Baldy, who had been the ringleader for their escape, was now in tow, Billy willingly followed along as they walked four miles back to camp.

After Bunk had finally recovered from the strenuous "Billy and Baldy roundup," the two men went back to the fossil bed to work on their specimen. With the passing of every hour, Bunk was more and more pleased with the large reptile. Late that afternoon it began to rain, forcing them back to camp. The next morning the ground was still too wet to work on the fossil, so they hunted birds with considerable success. After "making-up" the birds, they drove the wagon to town for mail and groceries. They received a letter from Martin advising them that he would arrive early in the morning on July 10. That news was well received, as Bunk and Theo were eager to get away from the grind of fieldwork. To celebrate the prospect of returning to Lawrence, they bought a piece of steak as a special treat for supper that night. Much to their disappointment, it was round steak that turned out to be tough and tasteless. The meal reminded them how much they had missed home cooking.

With the news of Martin's pending arrival only a few days away, they turned their attention to uncovering as much of the large fossil as they could before he arrived. They also had to gather up all of the bird and mammal specimens and ship them back to Lawrence before they departed for home. Back at the dig site, approximately twenty feet of the reptile had been uncovered. The vertebrae in the tail area were still about five inches in diameter, about the same as those in the rib area, indicating that there was still much more sea serpent to excavate. On July 8, they had cleared away enough of the soil to expose a front flipper bone, a hind appendage, and more vertebrae. Given the size of the bones and where they lay in the ground, Bunk calculated they were only half finished exposing the entire fossil, meaning the sea serpent could possibly be forty feet in length.

The day before Bunk and Theo were scheduled to leave Wallace County, they went to visit Allaman one more time. When they arrived, Bunk learned that the old man had been entertaining company all that day, which concerned him that Allaman might be all "talked out." However, when Bunk suggested that they take him in their wagon to see the fossil, "that seemed to be the proper spring" and Allaman jumped at the chance.

As they drove out to the site, it became apparent that Allaman was not all talked out. He told stories about the old days, including, as Bunk recorded it, "early history and animal traits of the gray wolf, coyote, Indians and English noblemen." Allaman explained how the early cattlemen tried to eradicate the gray wolf population in western Kansas because they were such a threat to their livestock. To do that they sprinkled strychnine on the carcasses of dead buffalo so that the gray wolves would eat it and die. Bunk would later recall Allaman's story when examining the outcomes of certain human activity in one of his few scholarly writings, about the demise of the kit fox. The small fox cohabitated with the wolves on the high plains at that time and because it was so quick it would often get to the poisoned bison carcass before the wolves. As a result of that practice by the ranchers, both species were eliminated from the area, even though the kit fox was no threat to the cattle.

Allaman also told stories about how smart and resourceful the coyotes were. He told them about a coyote with a broken front leg. By holding the broken leg in his mouth, the coyote was able to run without difficulty on the other three. Another story concerned ranchers trying to trap a coyote. A chicken was enclosed inside a barrel, which was set on the ground within a circle of leg traps. Although the traps were sprayed to remove the scent of man, they could not fool the coyote. The crafty animal maneuvered the barrel around the traps, then opened it and got the chicken.

When they arrived at the fossil site, Allaman carefully examined Bunk's fossil. Even though Allaman acknowledged that he "had shipped tons of saurian bones" from around that area when working with Williston and all of the other fossil men, he was perplexed because he had never seen anything quite like this one. According to Bunk's notes, Allaman "believes there will be a big dispute by fossil men over this species of the specimen, but says it is a very valuable find."

On July 10, the Union Pacific train pulled into the town of Wallace at five o'clock in the morning carrying Martin and three other men in his fossil collecting party. The train arrived slightly earlier than Bunk had expected; as the wagon neared the train station, he could see his good friend "Martin standing on a box on a truck waving his arms like a wild man."

Bunk was excited to show off the fossil remains, so first things first. After all of the fossil team's equipment was loaded into the wagon, he drove Martin to the fossil bed. Martin agreed with Bunk's assessment and "said it was at least forty feet long and a new species," which pleased Bunk to no end. The rest of the day was occupied with getting ready to take the 5:30 p.m.

Union Pacific back to Lawrence, where they would arrive at seven o'clock the next morning for a much-deserved rest and a chance to share the story of his discovery with his colleagues at the museum.

After Bunk returned to Lawrence, he appreciated regular warm baths and Clara's hot meals. A few days later Bunk ran into his old friend Pug Saunders on the street. Bunk anxiously told him about the giant sea serpent fossil he had found in western Kansas. As they visited, Saunders asked Bunk if he had heard about Dyche's run-in with some of the local fishermen. As Bunk had not, Saunders told him about the incident. On May 25 Dyche, in his role of game warden, arrested and charged a group of eleven fishermen illegally net fishing down by the dam on the Kaw River and hauled them through the city to the courthouse. The conflict between commercial fishermen trying to make a living and a state law making it difficult for them had come to a head and put Saunders in a difficult situation. Dyche had recently hired Saunders as a deputy game warden for Douglas County and now Dyche had arrested and humiliated some of Saunders' closest friends. Although Dyche later acknowledged that he was sorry to do that because many of them were his friends also, the relationship between the two men continued to erode. Saunders understood why hunting laws existed, but as a lover of hunting and fishing and loyal to his friends he was sympathetic to those who rallied against the laws.

It did not take long after Bunk's return for the citizens of Lawrence to call on their hometown animal expert. A week later Bunk received a telephone call from the local wholesale grocer in east Lawrence. The grocer explained that he needed assistance dealing with a snake. As the local newspaper would report, the worried grocer was confronted with a snake in a crate of bananas that had recently arrived from South America. Bunk responded immediately, only to find a tiny, harmless boa constrictor cowering in the corner of the packing box. The snake presented no actual danger, but it caused a considerable stir for the grocer and his employees. Bunk captured the boa and added it to the university's collection.

On August 28, 1911, nearly seven weeks after they arrived in Lawrence, Bunk and Theo returned to Wallace to resume their collecting trip. Martin picked them up at the train station and drove them back to his camp, which was near Bunk and Theo's old campsite. Rather than visit the fossil dig site, Martin updated Bunk and Theo on his activities and explained that in addition to excavating the giant sea serpent, they had taken the wagon to Colorado for fossil collecting.

Martin and his men spent the rest of the day packing for their trip home. At five thirty that evening they boarded the eastbound Union Pacific and headed back to Lawrence. With Martin's departure, Bunk and Theo began to regroup so they could leave town the next day to resume their bird- and mammal-collecting expedition.

The next day, Bunk wanted to take one more look at the site of his large fossil. After a short walk, he arrived at the dig site. Although Martin had taken up the fossil they had originally exposed, Bunk was surprised to see that Martin had not worked on the section of the ledge that would presumably be home to the tail section of the sea serpent. While understanding that it was common for fossilized skeletons to be incomplete due to the shifting of ground over the millions of years or because an animal's bones can be dispersed shortly after its demise by predators, Bunk, however, wanted to make certain they weren't missing important fossilized remains of the sea serpent. He began to dig and poked around the unexcavated ledge. After several minutes he confirmed his suspicion: a considerable amount of the sea serpent fossil still resided in the rocky shelf. When Bunk returned to camp, he informed Theo that he could stop packing the wagon; they weren't leaving until they unearthed the remainder of the sea serpent.

Bunk and Theo gathered up some tools and returned to the fossil site. After digging around the rocky outcropping for most of the afternoon, they were able to uncover seventeen new tail vertebrae. Bunk could not tell whether Martin had overlooked the additional bones or had concluded that they were safe from other fossil collectors since they were still unexcavated. In either event, now that Bunk had exposed the fossils, he could not leave them for possible erosion from the elements or the mischievous hands of competing fossil collectors. This was too important a specimen, so Bunk made the decision that he and Theo needed to finish the excavation. The resumption of their collecting birds and mammals would simply have to wait a few days while they finished excavating the large fossil. Considering the physical exertion required to complete the excavation versus the effort associated with hunting for birds, Bunk and Theo often shared their second thoughts about that decision over the next week.

The next day Theo began digging at the site while Bunk drove the wagon to town for plaster and burlap so they could prepare "field jackets" to wrap

Clark (hired hand) and Theo (wearing gloves) at dig site for the fossil of the forty-five-foot mosasaur (University of Kansas Natural History Museum records, University Archives, RG 33/0 Photographs, Kenneth Spencer Research Library, University of Kansas Libraries)

and immobilize the fragile fossilized bones. They also needed lumber for boxes, a shovel, and horse feed. On the way back, he stopped at Pinnell's for a pickaxe and at Allaman's he borrowed a shovel. While they sat exhausted on the rock ledge after they had finished digging all afternoon, they reflected proudly that they had unearthed a total of twenty-one vertebrae and had "a hole in the side of the bank big enough to build a dug out," referring to the homes early setters hollowed out of hillsides.

Midway through the next morning, Bunk decided that they needed an extra pair of hands for digging. He went to Wallace and asked the owner of the grocery store whether there was anyone he could hire who could help dig. Later that afternoon, the grocer showed up at camp with a man named Clark, who was willing to work for twenty cents an hour. Although the digging went faster with the extra hands, it was still exhausting work.

On the third day of excavation, they finally had enough of the tail section exposed so they could begin making the plaster and burlap "field jacket" to safely remove the bone segments from the rock. Although Bunk had never made one, he had seen many in the museum. He thought it couldn't be that different from his work with plaster when he and Theo built the scenery in

Plaster field jacket surrounding a portion of the fossil prior to its removal in 1911 (University of Kansas Natural History Museum records, University Archives, RG 33/0 Photographs, Kenneth Spencer Research Library, University of Kansas Libraries)

the panorama. He basically understood the process and was confident that it would work. While the plaster dried they dug and picked at the rock underneath the fossil to undermine the field jacket for removal. The next day, when the plastered section was removed from its location, it was intact, but when they turned it over part of it fell away. Although they were disappointed, it was no surprise, as the soil was very flaky and unstable. Any fragments that could not be taken in the slab of plaster were wrapped in burlap to avoid further damage to these fragile artifacts.

The process of uncovering portions of the fossil, creating a plaster slab, undermining the section, and removing the section continued until September 4, their last day at the fossil site. In the afternoon, they loaded all of the plaster sections and the loose fragments that were in burlap onto the wagon and returned to camp. Once Bunk and Theo finished constructing shipping crates, they carefully packed the last of the fossilized bones of the giant sea serpent. On the morning of September 5 they hoisted the boxes of artifacts onto their wagon and drove them to the train station in Wallace. According to Bunk, with the crate safely loaded onto the train, "that was quite a load off our minds." They returned to their campsite to pack everything up to move north the next morning. On September 6 they left their camp on the Smoky Hill River. They veered a half a mile out of their way to say goodbye to Mr. and Mrs. Allaman, who offered them seven musk melons and a couple of quarts of Burbank plums and wished them well.

Bunk and Theo planned to hunt their way back to Lawrence. They would drive the wagon to the northwest corner of the state and turn east to drive along what today is State Highway 32. They had prearranged an extended hunting stay in Concordia, Kansas, on the Republican River, a little more than halfway across the state. Those arrangements were made through David Horkman, the Lawrence photographer who had taken the 1906 panoramic photograph of the *Panorama of North American Mammals*. His brother lived in Concordia and made plans for them to hunt. By October 19 the weather turned cold and the birds all but disappeared, so the expedition concluded. Bunk took the train from Concordia back to Lawrence and Theo would drive the horses and wagon. A young man named Frank Ward, a nephew of David Horkman's brother, accompanied Theo back to Lawrence, where he was going to work in the photograpy business.

By the end of October Bunk, Theo, and all the boxes of birds, mammals, and fossilized sea serpent bones were back in Lawrence. Bunk and Theo had completed their successful collecting trip, circling much of the perimeter of Kansas. After riding for over a thousand miles on the hard bench of a wagon, Bunk was proud of their bounty. The trip added nearly 900 birds to the museum, plus 31 mammals and 12 sets of bird eggs. In addition, they brought back the stomach contents of all mammals, hawks, and owls killed; plus, there were 8 snakes, 125 turtles, clam shells from different streams, fossil oyster shells, and 35 skulls of coyotes, badgers, wildcats, raccoons, and prairie dogs. But the cherry on the ice cream sundae was the skeleton of the enormous fossilized sea serpent he had found in the Niobrara Chalk in Wallace County. Their entire budget for the nearly six-month trip was less than $110. That included food, a gasoline stove and fuel, shipping materials and postage, an occasional ten-cent cigar, the expenses of feeding and shoeing the horses, and train fare for both Bunk and Theo from Wallace to Lawrence and back. The amount was over 20 percent of the total funding for the entire Kansas Biological Survey that year, and there were nearly a dozen other collecting trips that emanated from KU that summer, though none were as extensive as Bunk's. Incidentally, it was fortunate that the board of regents

Artist's rendering of Bunker's mosasaur (painting by Daniel Varner, used with permission of Mike Everhart, Oceans of Kansas Paleontology)

approved the funding of the Kansas Biological Survey, because if the request had been denied Bunk would have cut the trip short and never found the sea serpent fossil.

Bunk was glad to be home, but he was already planning a collecting trip for the following year. In spite of the rugged hardships of camp life, Bunk always looked back fondly on his collecting trips. He always treasured viewing the majestic expanses of the Kansas landscape, and he loved both the quiet solitude of sleeping under the stars on the prairie and the nervous anticipation of sneaking closer to a bird for a better shot. He remembered the thrill of discovering something new, especially something as significant as the fossil of the giant sea serpent that lived millions of years ago.

Fieldwork was a simpler life, and something Bunk especially enjoyed sharing with Theo. Although there were many years' difference in their ages, they seemed more like partners than supervisor and assistant. The ultimate responsibility for the outcome of their collecting trip fell to Bunk, but they shared all duties equally. Both set up tents, cooked meals, washed dishes, and shared in the hunting and skinning of the collected specimens. Theo tended the horses and reloaded shotgun shells more often, but Bunk set and collected the traps and wrote the field notes. Most importantly, they

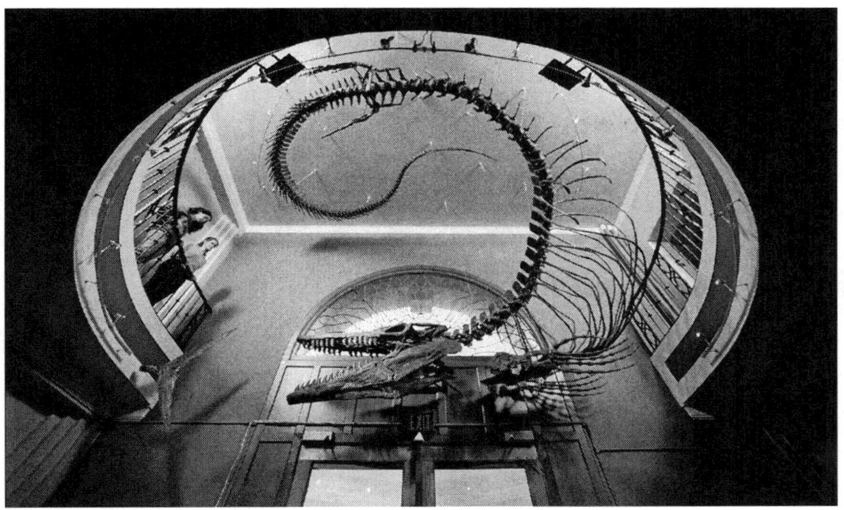

Replica of Bunker's mosasaur hanging over the entryway at Dyche Hall (© 2003 University of Kansas/Marketing Communications/Doug Koch)

always got along and truly enjoyed each other's company. Even when they spent long days digging the enormous fossil under the hot sun or driving the horses into the teeth of the dust storm, there was never any complaining or bickering.

When the crates of the giant sea serpent fossil arrived in Lawrence Bunk saw that Allaman had correctly predicted the fossil's significance. Initially the competing theories of whether it was a plesiosaur (Allaman's contention) or a mosasaur or something entirely new were based primarily on the sheer size of the specimen. Previously discovered mosasaurs were much smaller than the one Bunk excavated, and most paleontologists thought only plesiosaurs grew that large. In the years that followed, the fossil remained boxed in storage because it was too large to be displayed in the existing exhibition space in Dyche Hall. Martin and the faculty eventually agreed that it was a mosasaur, but it would take decades before they would determine the genus and species.

Allaman's admonition that "there will be a big dispute" regarding this fossil was not limited to determining the species. Bunk left a handwritten notation on the back of one of the photos taken of the dig site indicating that the absentee landowner contested KU's right to the fossil, saying that permission had not been granted. According to Bunk's note, a lawsuit ensued, and the "Supreme Court" ruled that the foreman had given permission on behalf of Pinnell, and thus the remains of the giant sea serpent belonged to the KU Natural History Museum.

In the late 1960s, when enough of the fossil had been revealed, paleontologists concluded that Bunk's mosasaur was a *Tylosaurus proriger.* Finally, in 1999 a replica was cast and suspended from the ceiling above the main entrance of the museum at Dyche Hall. Today all visitors are greeted by Bunk's giant sea serpent, coming face-to-face with one of the largest and most ferocious sea serpents that once dominated the shallow ocean of Kansas millions of years ago. In 2014, more than one hundred years after Bunk returned to the University of Kansas bearing wooden crates of the fossilized bones of the sea serpent, the governor of Kansas signed into law a resolution making *Tylosaurus* the state's official marine fossil.

CHAPTER ELEVEN

Period of Great Loss and New Responsibilities

On September 20, 1912, Theo Rocklund boarded the train in Lawrence for western Kansas. Theo was traveling alone and headed to Pendennis, a small unincorporated community in Lane County, to meet Handel Martin, who had been hunting fossils. Theo would take over the team and wagon, so Martin could take the train back to Lawrence. Martin asked Bunk to be present at a zoology department meeting. Once the meeting was concluded, Bunk would take the train to join his hunting partner. As the train rumbled westward, Theo worried about the health of their two reliable steeds, Billy and Baldy. In a letter dated September 11, 1912, Martin had advised Bunk and Theo about a serious horse disease that was rampant in western Kansas. In the previous week twenty horses in Lane County, within a few miles of where he was camped, had died from the sickness, all rather quickly. Knowing the importance of the two horses to the success of their collection activities that autumn, Theo was justifiably worried.

The department meeting was scheduled to elect an interim chair of the zoology department because Clarence McClung had recently announced that he was leaving the university due to the financial constraints put on the university by the legislature. Martin and Bunk always kept a watchful eye on any changes in the faculty to make sure nothing occurred that would negatively impact their influence and independent control over their respective collections. In this particular situation, Martin's concerns revolved around who ultimately would be chosen as the new chair and how that selection might affect who would become the curator of the fossil vertebrate collection. The chair of the department and the curator of fossils was a dual appointment that McClung had inherited when Samuel Wendell Williston left the university, but there was no guarantee that one person would hold both positions going forward. As soon as the department meeting concluded,

Bunk hopped aboard the next train west to join Theo for the second chapter of their extended survey of Kansas birds and mammals.

Like the previous fall, Bunk and Theo planned to hunt for birds and mammals until the cold weather drove them home. They would focus on Lane, Gove, and Trego Counties, a decision that was somewhat predicated on Martin's fossil collecting expedition. Because fossils were so common in those counties and they shared the same team and wagon, it made sense for Bunk and Theo to hunt there also. The area included several tributaries of the Smoky Hill River and Walnut Creek, a hundred-mile-long tributary of the Arkansas River, and Bunk felt that as long as there was a source of water for the birds, the hunters would do just fine.

As the autumn progressed, with every cold front from the north came a new flock of migratory birds. With the prevalence of birds that year, their days were long and filled with shooting and skinning. Bunk and Theo watched autumn turn into winter with a snowfall on October 31. When the snow stopped around midday, they resumed hunting. The snow made the legs of their trousers so wet that they had to work the rest of the afternoon in camp to let them dry out. Finally, they decided that November 1 was their last day to hunt.

When Bunk and Theo wearily returned to Lawrence in mid-November, Bunk proudly reported to Clara that their bounty from this trip included over seven hundred bird skins, a box of Puerto Rican cigars, and a "brush gun." Embarrassed about what happened, Bunk, who was normally a very careful hunter, reluctantly told Clara the story about the gun. On October 23 he was hunting on the ranch of Dr. Jones, south of WaKeeney, Kansas, when he shot a duck. Although he knocked the bird down, it was not a fatal shot and before he could get to the duck, it scampered to cover under some Russian thistle. Bunk poked around in the bush with his shotgun and inadvertently got mud stuck in the muzzle. Unfortunately, he forgot to check the barrel before his next shot and was startled when the shot blew off the end of the left barrel and damaged the rib of both barrels for about six inches. Bunk realized that luckily only his pride was hurt with this mishap. When Dr. Jones' son cut off the end of the damaged barrel, Bunk sheepishly admitted that at least it would make a good "brush gun." As for the cigars, that was a happier story—they were a gift. Near the end of their trip, they went to town for mail and Bunk found a package addressed to him from Doc Wetmore. When he opened it he found a box of Puerto Rican cigars. Wetmore had been married earlier in the month and sent the cigars to Bunk in celebration. Bunk was sorry he had been unable to attend Wetmore's wedding, but joked to Theo that if anyone would understand that he was missing it because he was collecting birds, it would be Doc. The cigars were also in appreciation of Bunk for introducing Wetmore to his wife-to-be. For years Bunk had been good friends with the neighborhood grocer, Scott Holloway, who had a tennis court in his backyard. On one occasion, Bunk invited Wetmore to join him in playing tennis and Wetmore met the grocer's oldest daughter, whom he married in the summer of 1912.

Despite not returning with anything as exotic as a giant sea serpent fossil, Bunk felt that the 1912 survey trip contributed immensely to the bird collection. For months he worked on identifying and cataloging the new bird skins. He engaged the help of his capable assistant, Arta Briggs, and a student by the name of Dix Teachenor. Teachenor grew up in Kansas City, Missouri, and came to KU to study liberal arts and play tennis. While at KU he discovered his love of nature and eventually became one of Bunk's boys. From his experiences at the cabin he developed into a proficient collector. However, unlike most of Bunk's boys who went on to careers in science, when Dix graduated he returned to Kansas City and began a successful career in the insurance industry. Despite the fact that he did not follow a career in science, his experiences collecting with Bunk led to a lifelong interest in ornithology. Because of his passion for birds, he built a specimen laboratory in his home. For over forty years Dix collected in the Kansas City region, often contributing specimens, mostly bird eggs, to the KU Natural History Museum. Today nearly eight hundred specimens in the ornithological collection are attributed to him.

Although Bunk's scientific writing was limited during his career, those two major western Kansas expeditions led to a published article that updated Frank Snow's comprehensive 1903 survey of birds. In June of 1913, the *Kansas University Science Bulletin* (volume 7, number 5) published Bunk's article "The Birds of Kansas." As usual, Bunk gave credit to his hunting and collecting partner, Theodore Rocklund, and to Miss Briggs, who diligently rechecked the entire collection to determine any additions and deletions from Snow's edition. The final count was over 1,000 birds collected in 1911 and 700 in 1912. In all, he sighted or collected 379 different species or subspecies of birds.

Bunk was still basking in the glow of his article when out of the blue, Theo approached him about an opportunity for the two of them to acquire a Hup-

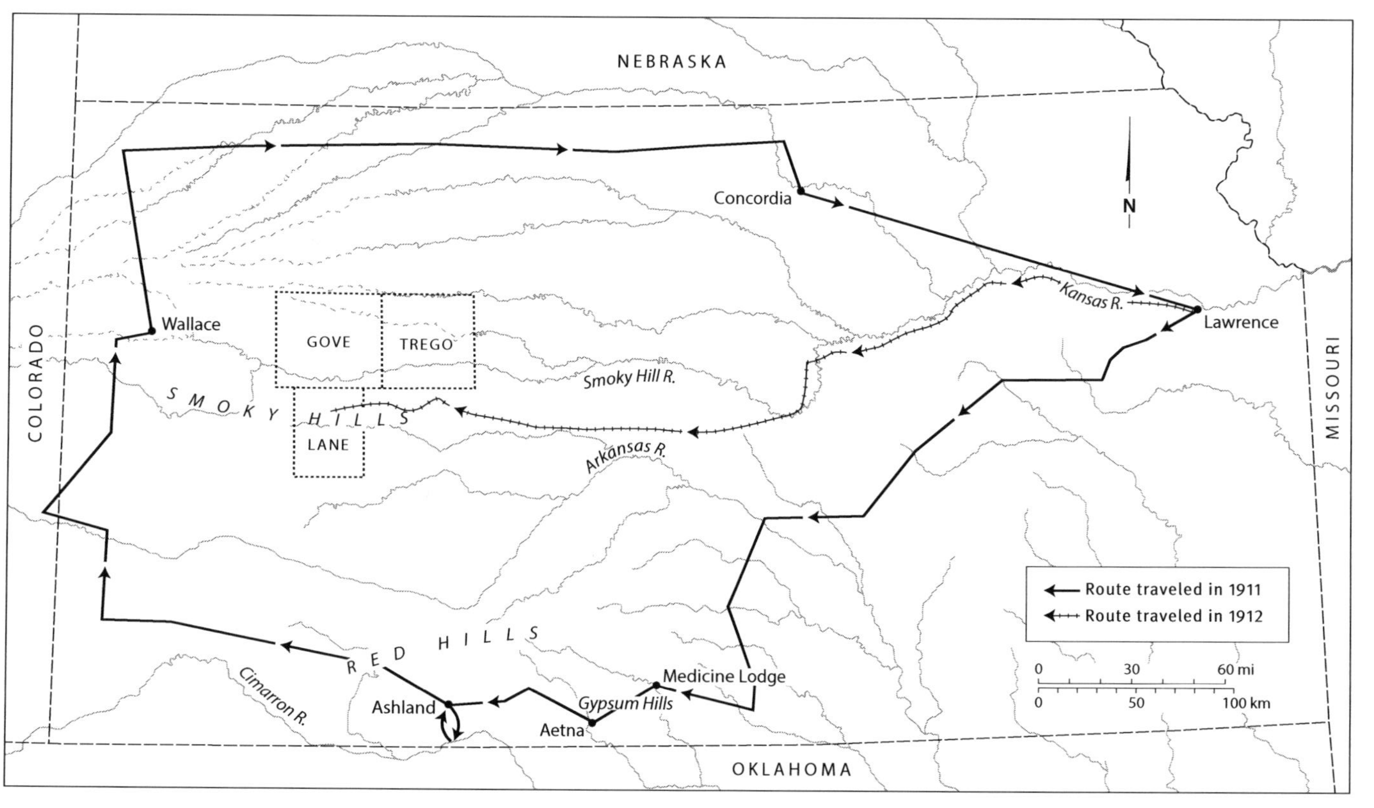

Route of 1911 and 1912 collecting trips

mobile automobile dealership. Throughout Bunk's career at the museum, he always struggled to make ends meet on his meager salary, so the prospect of earning some extra money was worth considering. At first Bunk was skeptical about the idea, especially since he knew so little about the automobile business. Theo was excited about the prospect, thinking that automobiles would be the wave of the future. Eventually Bunk agreed to join his friend in the venture. As it turned out, the cars did not sell themselves, and with shy Bunk being uncomfortable selling, they eventually exited the business. The two of them sold two cars: one to themselves and one to a man named Walter Kibbie. When Kibbie passed away several years later in California, he willed the car to Bunk because "his customer service was so good." Appreciating the thoughtfulness of his former customer, Bunk took the train to Riverside, California, and drove the 1914 Hupmobile back to Lawrence.

Despite the bad luck of the auto business, the flow of gifted students continued and in 1913 two new young men came under Bunk's tutelage. Victor H. Housholder, a native of Columbus, Kansas, was tall, athletic, and good-looking. He had enrolled several years earlier in the school of engineering, but had always loved hunting and the outdoors. Beginning that semester, he earned a spot on the collecting trips to Bunk's cabin and frequented those weekend hunts for two years, before earning his degree in 1915. He continued to work and study at KU until he took a job in Wyoming working for a Tulsa Oil syndicate. As a first senior lieutenant in World War I, he served under president-to-be Captain Harry S. Truman. During fierce fighting in France, Housholder rescued Truman from under his fallen horse. His friendship with Truman continued throughout his lifetime and he even spent a weekend at the White House during Truman's presidency. Housholder enjoyed a highly successful engineering career in Arizona and was named engineer of the year by the American Society of Civil Engineers in 1957.

The second outstanding student was Remington Kellogg, who excelled as a wildlife collector even before attending Kansas University. Growing up in Kansas City, Missouri, he had looked for a college with an excellent reputation for natural history and selected KU. Initially Kellogg studied entomology, but later changed to mammology. During his freshman year he developed a close friendship with Alexander Wetmore. From 1913 to 1916 he worked as a student assistant with Bunk, during which time Kellogg published his first scientific paper, in 1914, comparing a series of wood rats from eastern and western Kansas that had been collected by Charles H. Sternberg in 1895, R. C. Lindsey in 1910, and Bunk and Rocklund in 1912. When a pro-

Handel Martin (*left*) and Bunk at Bunk's cabin circa 1913 (University of Kansas Natural History Museum records, University Archives, RG 41/0 Photographs, Kenneth Spencer Research Library, University of Kansas Libraries)

fessor unexpectedly died during Kellogg's senior year, KU recruited him to take over the professor's class in ornithology. Kellogg received his undergraduate degree in 1915 and his master's degree in 1916. The subsequent fall he entered the University of California, Berkeley, to study zoology. Eventually he became interested in vertebrate paleontology, specializing in marine mammals. Kellogg's graduate studies were interrupted by World War I. In

Bunk and Theo Rocklund at Bunk's cabin circa 1913 (University of Kansas Natural History Museum records, University Archives, RG 41/0 Photographs, Kenneth Spencer Research Library, University of Kansas Libraries)

December of 1917 he enlisted in the army and was ordered to France, where he was assigned to the command of a general field naturalist and worked on controlling rats in the trenches. During his time in France, he was able to collect birds and small mammals, which he sent back to both Berkeley and Kansas University.

With the new students came frequent jaunts to hunt at the cabin. Wetmore, Kellogg, and Housholder were usually included on those weekends, and often Martin and Rocklund joined them as well.

For Bunk, 1915 was a transformative year. His comprehensive cataloging system for birds and mammals operated quite smoothly, at least for the new specimens; however, the entries for specimens collected prior to 1910 still contained large gaps in the information that Bunk wanted to include in the permanent record. To fill in these gaps, Bunk often had to search through collectors' old field notes or pull the specimen out of storage to find the missing information. This was so time-consuming that Bunk often sought out Dyche to fill in the gaps. Because Dyche was still living in Lawrence while he worked at the state capitol, Bunk had to either write to him in Topeka or catch him whenever he passed through the museum. Bunk's persistence in getting information from Dyche turned out to be critical to the collection. While lobbying in Topeka on January 14, 1915, Dyche became ill, and a few days later he was hospitalized. On January 20 he died, at age fifty-seven, due to a recently discovered heart problem. Dyche's sudden passing shocked the entire university community. To memorialize him, the KU Natural History Museum was renamed Dyche Museum. For Bunk the loss was profound. Even with their personality differences, Dyche had given him the chance to work at a world-class museum with many talented professionals, an opportunity that altered his life and made Bunk forever grateful.

With his death the question arose of who would take Dyche's place as the curator of the collection of birds and mammals. Bunk who was forty-five years old, had a wealth of museum experience, and currently serving as assistant curator in charge, hoped the chancellor would appoint him curator. To plead the case for his appointment, Bunk wrote a letter to Chancellor Frank Strong: "I ask if twenty years of active work in museums with always the best interest of the department at heart is not the training to make the best head? Or would you handicap us by placing some well meaning higher educated man over us, who knows absolutely nothing of the care and management and growth of a museum?"

Before receiving a reply from the chancellor, Bunk heard rumors that someone else within the university was trying to gain control of his collection of birds and mammals. In 1913 Bennet M. Allen succeeded McClung as head of the department of zoology and curator of the vertebrate paleontology collection. When he learned of Dyche's death, he wrote a letter to Strong proposing the consolidation of Bunk's recent vertebrate collection with vertebrate paleontology. In addition, he suggested that the combined collection be under the control of the zoology department and that he, Allen, serve as curator.

Infuriated by the news of this assault, Bunk went on the offensive in the form of a carefully worded letter to Chancellor Strong stressing the importance of the natural history museum remaining independent. In Bunk's view, the museum served the entire university, not just one department. He argued that Allen's proposed organization would give preference to zoology, while disadvantaging all other departments.

In that same letter, he took the opportunity to once again make his case for the position of curator despite not having a college degree. He wrote, "I wish to say that in my mind it is not wholly necessary that a man have a degree to properly manage a museum. Dr. Skiff, the director of the great Field Museum in Chicago . . . [received] his degree . . . as an honor for his success as a museum builder . . . [and] many others hold their position through their ability rather than a degree." Not having a degree made his appointment as curator an uphill battle, but his plea demonstrated his determination and passion for the museum. Considering that those comments came from such a quiet and unassuming man, his actions displayed commendable courage.

Fortunately, Bunk found a solid ally in this turf war. McClung gladly recommended to the chancellor that Bunk succeed Dyche as curator, stressing that he "did not want to do anything that would help perpetuate the old order." To McClung, "the old order" referred to Dyche and his emphasis on collecting for exhibit and display rather than for instruction and research. McClung felt that under Bunk's direction the collection would better support a balance of exhibition, teaching, and research.

When Strong finally replied to Bunk on November 11, he expressed his sincere hope that Bunk would remain in charge of his collection, but only offered Bunk the title of Assistant Curator in Charge of Recent Vertebrates. In spite of pleas by both Bunk and McClung, Strong and the board of regents refused to consider his appointment without the qualifier of "assistant." By 1915 times had changed at universities and it was considered unbecoming for a curator of a major collection to have only a grammar school education, especially at an institution that was in the business of bestowing academic degrees. Despite his extensive scientific knowledge and establishment of a sophisticated and thorough catalog system for the specimens in the collection, Bunk still lacked a college degree. In the end, while the chancellor considered Bunk a valuable and capable "museum man" who should continue to exert direct control over the collection, he could not grant Bunk the title he wanted most. Although Bunk was disappointed, at least Allen's attempt to control a collection had been successfully rebuffed — though not for the last time.

To help soften the blow, the chancellor raised Bunk's salary and gave him control, at least on paper, over any future candidate for curator of his collection. According to Strong's letter, if at any time in the future the university wanted to bring in someone to be curator of the collection, it would be done only with Bunk's "accord and cooperation." In essence, it seemed the chancellor was offering Bunk a kind of right of first refusal for any successor. However, since Bunk still reported to the chancellor, Bunk held a weak hand for any future attempts to install a new curator. In the end, Bunk was satisfied with the terms and continued on at the museum with the title of Assistant Curator in Charge of Birds and Mammals.

With his new title and responsibilities, Bunk entered 1916 full of optimism. It was another year that would prove eventful. That spring, Kellogg graduated and left Lawrence for graduate school at Berkeley. In the late fall, there were reports of sightings of American goshawks in eastern Kansas. One of the largest of the hawk family, they usually preferred a more northern climate. Feeling that it was important for someone to report the presence of a newly sighted bird in the state, Bunk submitted a short article to the American Ornithologists' Union, one of the oldest and most respected ornithology organizations in the country and publisher of two widely read journals. The article was published in the society's oldest publication, the *Auk*, in January 1917.

In 1916 the national debate over protecting birds continued to boil. Previously the only protections offered to birds, fish, and other wildlife originated in individual states, which resulted in laws that lacked standardization. Considering that the same birds may fly over multiple states, as well as international boundaries, Congress in 1916 began writing a uniform law that would eventually become the Migratory Bird Treaty of 1918 between the United States and Canada.

A wide variety of competing interests lobbied to shape the legislation, including hunters, who wanted to make sure they had a supply of game birds; farmers, who wanted protection for birds that ate crop-damaging insects; and conservationists, who wanted to ensure the survival of species. There also were those who believed that the conservationists were overreacting, like they did in the previous century when they wanted to protect the passenger pigeon. Although the species had once been plentiful across the eastern half of the United States when the Europeans first settled in America, by the 1880s passenger pigeon populations had declined significantly as a consequence of being widely hunted as a source of free food. Those opposing the legislation claimed that the birds, like the bison, were so abundant that they did not need protection. Sadly, the last passenger pigeon eventually died in captivity in a zoo in Cincinnati in 1914.

Ironically, Bunk had a story involving a passenger pigeon. As a young boy back in Mendota, he spied a passenger pigeon perched in a tree. Bunk shot the bird and delivered it to Banning to mount. Once Bunk was aware that the passenger pigeon had finally become officially extinct, he wrote Banning to ask if he could get the mounted pigeon for the museum. Banning replied that it would cost Bunk one hundred dollars. Once again, Bunk did not have the money.

The new laws of 1916 prohibited the hunting, killing, capturing, possession, sale, transportation, and exportation of migratory birds and their feathers, eggs, and nests. Certain game birds were excepted, but only during their established hunting seasons. In addition, there was no protection for birds that had been transplanted from another country, such as starlings and English sparrows.

In the fall of 1916 Bunk joined the national discussions surrounding the war on English sparrows, also known as the house sparrow. The birds were first transported to the United States from England in 1852 to combat the cankerworm infestation attacking one of the country's favorite shade trees, the American elm. English sparrows quickly acclimated to their new environment and thrived. They were feisty little birds, compared to some native species, and developed a reputation as a pest. Some ornithologists began to reconsider their value, fearing that they were competing too successfully with native sparrows and other songbirds. The noted naturalist and nature essayist John Burroughs announced that the alien sparrow should be eliminated immediately. The League of American Sportsmen joined the chorus, asking every state to fund the eradication of the intrusive species because they ate grain from the farmer's field, aggressively stole the nests of other songbirds, and drove away game birds that the association wanted to hunt. In addition, the Department of Agriculture issued a bulletin that declared the English sparrow a pest because it ate valuable seed crops.

As the face of natural sciences for the state, the KU Natural History Museum often became an advocate for wildlife. As soon as Bunk got wind of the outcry to eradicate the English sparrow, he decided to act. On November 27, 1916, Bunk wrote to the *Kansas City Star*, hoping the newspaper would defend the species. His letter reasoned that even though the English sparrow ate grain, it also ate harmful insects, which native sparrows did not. He explained that the problem of the house sparrow occupying man-made birdhouses intended for more popular songbirds was due to the fact that they were not a migratory bird. That is, unlike their native counterparts, house sparrows stayed put and foraged over the winter, and as a result were the first birds to occupy existing birdhouses. He further explained that birdhouse builders could avoid this problem if they just made the entrance no larger than a "silver quarter." He pointed out that even though house sparrows ate grain, more often than not they took the grain off the wagon or from bins rather than out of the field because of their tendency to locate near farm buildings. This predilection for wanting to be near structures also made them more likely to be in town than on the farm. Also, they frequently feasted on the seeds of weeds, the eradication of which was a good thing. Finally, he pointed out that many other birds eat grain but don't have a reputation for being a pest. For instance, swarms of doves and grackles "darken the sun on their way to the grain fields," but were protected. Bunk warned against doing something rash, impetuous, and possibly irreversible to a species of birds just because of their propensity to inhabit the homes of preferred songbirds.

In any event, despite the efforts of the League of American Sportsmen to mobilize "all the college men and school pupils and Boy Scouts," as soon as the war began it suddenly came to a halt. It was found that birds other than so-called English sparrows were being largely destroyed by mistake. This led naturally to the question: If birds other than English sparrows were mistaken for sparrows by the sparrow hunters, was it not likely that birds other than English sparrows had been responsible for the original acts that caused the uproar? Bunk knew that a bona fide bird watcher could easily distinguish a native sparrow from an English sparrow, but trying to mobilize amateurs resulted in chaos and needless bird carnage. It was like the old joke about the non-ornithologist who claims to know all about birds: "There are the little brown ones, the middle-sized brown ones, and the large brown ones."

In December, Bunk received an unexpected letter from the American Ornithologists' Union (AOU). The letter came from Mr. John H. Sage, the secretary of the organization, advising Bunk that at the Thirty-Fourth Stated Meeting of the AOU held the previous month on November 13, 1916, in Philadelphia, he had been elected as an associate of the AOU. Bunk held the organization in high esteem and deeply valued the recognition from his peers. The public notice of all new associates of the organization was published in the *Auk* in January 1917.

Over the years Bunk tried to stay in touch with his "boys" after they left KU. They thought of each other as family, and whenever they were near Lawrence they went out of their way to stop by campus and visit with their old friend. In April 1917 Bunk wrote a newsy letter to Wetmore suggesting that he consider letting Bunk properly store and catalog his personal collection of birds. Bunk acknowledged that he had recently examined a large box of Wetmore's skins, which he described as "in fair shape" and showing no insect damage. Bunk thought that if they were properly cataloged, it would be easier for Wetmore to call for any series he wanted to study. To ensure that there was no confusion about the ownership of the skins, Bunk offered to note on each specimen tag that they were Wetmore's property.

Bunk also mentioned in his letter that the bird catalog was finally complete, and that a responsible young first-year student named Deane Malott had checked it against the collection, which now totaled 10,650 birds. Over three decades later, when Wetmore donated a large portion of his bird collection to the university, it was Malott who officially acknowledged receipt of the gracious and generous gift to the university, but in his capacity as the chancellor of the university.

Bunk closed his letter to Wetmore by asking how things were going in Washington, DC, with the war effort beginning. Although age prevented Bunk from serving in World War I, it did not preclude his good friend and companion Theodore Rocklund. Theo enlisted in the army in early 1917, shortly after the United States declared war. The following year, as the war continued to rage on, the annual influenza epidemic began spreading around the world. Unfortunately, the 1918 strain, known as the Spanish flu, had a mortality rate twenty times higher than the typical yearly flu. Unlike other flu strains, it presented the greatest risk to people between twenty and thirty-five years old, prime military age, and the presence of so many soldiers in Europe created an unprecedented catastrophe. According to one source, forty-three thousand American soldiers died of the flu that year — about the same number as those who died of wounds from the war.

During the fall of 1918 Theo found himself stationed in France. As a close friend of the entire Bunker family, Theo wrote Bunk's youngest daughter, Audrey, a letter and enclosed a small white handkerchief with purple flowers and a green vine embroidered around the edge and "souvenir de France" stitched in the center. A month after Armistice, Audrey wrote a thank-you note to Theo, saying, "I am so proud of it as I am the only girl I know who got anything from France." She ended her letter with, "We all hope you will be home soon and are so glad you did not get hurt." A few months later an envelope was delivered to the Bunker home. It was the one Audrey had sent to Theo and on the envelope were the words "return to sender," and next to the addressee, "deceased 4-29-19." In a sad twist of fate, Theo had died not from battle injuries but from the flu he contracted in the months following the Armistice. With the terrible news of Theo's death, Bunk was suddenly without his longtime hunting companion, coworker, and close friend. Theodore A. Rocklund died at the age of thirty-two in France in an unknown hospital far away from friends and family. Theo's death made such an impression on Audrey that she held on to the handkerchief and undelivered thank-you note throughout her entire life and passed them on in a box of old family photographs and other cherished mementos.

CHAPTER TWELVE

Journey to Alaska

Like Bunk's field trips in 1911 and 1912, many details of his trip to Alaska are well documented. All accounts of places, events, and quoted material mentioned in the next three chapters are taken directly from field notes and personal written accounts of the trip.

The remainder of 1919 after Theo died, Bunk regularly caught himself thinking about his close friend, especially when he was doing something he had done with Theo, like hunting at the family farm south of town. Walking through the quiet woods reminded him of the good times they had shared. Bunk kept his feelings to himself and never mentioned them to anyone, not even Clara, although she knew how deeply saddened Bunk was.

Fortunately, later that year an opportunity arose that would help Bunk take his mind off the loss of his friend: Bunk received an invitation to join a group of hunters going to the Kenai Peninsula of Alaska in search of mountain sheep, moose, and bear. From the beginning, the members of the expedition agreed that any animals collected would be contributed to the collection at KU, save a few trophy mounts for the hunters. Bunk's presence would ensure that the specimens would be safely cared for and preserved. The differences between big game hunting and scientific collecting gave Bunk pause, but it still represented the chance to grow the mammal collection. Bunk also hoped that visiting the wild frontier of Alaska, with its remote and spectacular beauty, just might provide a good tonic to help him overcome his recent loss of his close friend Theo.

Today two very different accounts of the hunting adventure have survived. The first was Bunk's contemporaneous record, written in his field notebook at the end of each day. His notes were straightforward and unembellished. For example, when he specified the members of the hunting party he simply listed their names. The second record of the trip came from one of the hunters, Raymond J. DeLano, a Kansas City attorney and real estate developer. By contrast, DeLano, whom everyone called Del, composed and typed his narrative of the Alaska trip after he returned home. His account includes considerable bravado and hyperbole. In flowery, flamboyant language, he described every detail, from scenery to the individual members of the hunting party, with a flourish. Together, the two dissimilar narratives of the trip present a broader coverage of events of the expedition than each individually, especially when DeLano and Bunk were separated. But even when they were together, their observations offered very different perspectives of the same events. Upon reading DeLano's account, Bunk declared that he wasn't sure they had been on the same trip. Regardless of the differences in the two versions, both convey and illustrate the excitement and danger of the expedition.

The organizer of the trip was Dr. John Outland, a noted surgeon who practiced in Kansas City, Missouri. Outland attended Kansas University in the late 1890s, where he distinguished himself playing football. When he later attended medical school at the University of Pennsylvania, he continued to play football, becoming both an All-American lineman in 1897 and an All-American halfback in 1898. In recognition of his success on the gridiron, in 1946 the Football Writers Association of America established the prestigious Outland Award to annually recognize the best interior lineman in the country. DeLano described "Doc John" as "voluminous of body and soul, the wheel-horse, the wit, the wag."

Outland invited two friends to join him on this big game hunt. One was Dr. James Masson, a native of Ontario, Canada, who received his medical schooling in Rochester, Minnesota, where he would later become the chief of the surgical staff of the Mayo Clinic. DeLano characterized Masson as "a round Falstaffian character who possessed panache and charmed the women." The second was Dr. Raymond Teal, who practiced medicine in Palco, Kansas, and was depicted by DeLano as friendly, with a "winning smile, a winning tenor voice, and a winning poker hand." The three physicians and DeLano represented the well-to-do sportsmen whose primary interest was high adventure and taking home trophy mounts. They were boisterous and full of confidence, and often their large personalities were the center of attention, especially when they gathered to sing songs and tell jokes. The final member of the party, George Plotts, a cameraman for Crosby Studios in Los Angeles, was invited so that he could create a vi-

The 1919 Alaska hunting party: *left to right*, DeLano, Masson, Outland, and Teal (Alaska State Library, Dr. John Outland, Alaska Big Game Hunt, 1919 Photo Collection, P425-17-25)

sual record of the excursion, as well as stock film footage of the magnificent scenery of Alaska for future use in Hollywood movies.

On August 9 the hunting party boarded the steamship SS *Jefferson* in Seattle to travel to Juneau. The trip gave everyone the opportunity to enjoy the majestic scenery of the Inland Passage. Once the ship left Vancouver Island, the beauty of the scenery captivated DeLano, who wrote, "The mountains were higher, waterfalls and cataracts everywhere tumbling down from the snow crevices above." The hunting party settled in aboard the ship, expecting their journey to be peaceful and uneventful. However, on the second day a thick fog engulfed their ship. Suddenly, from out of the mist came another large vessel, the *Prince Rupert*, heading directly toward the *Jefferson*. As the two ships approached one another, they signaled with whistles to let the other ship know how they were going to proceed. Apparently there was some confusion and the *Prince Rupert* suddenly turned in front of their ship. Fortunately, the *Jefferson* reversed its engines just in time to avert a collision. As the two ships passed each other with less than a ship's length between them, the close call gave Bunk quite a shock. As for the unflappable hunters, they resumed their cheerful conversations without missing a beat.

When the ship docked in Juneau, they had to change steamships before continuing to the Kenai Peninsula. With a few days before the departure of a second ship, they elected to take a side trip to Hoonah, a small scenic indigenous village fifty miles to the west on Chichagof Island. At Hoonah they could hunt, fish, and sightsee. To reach Hoonah they transferred their gear to the *Tasmania*, a small gasoline-powered launch owned by a young Norwegian.

The islands of the Inland Passage between Juneau and Hoonah offered shelter from the open sea, so when the small launch ventured out into the ocean, they encountered for the first time what Bunk described as fifty-foot-high waves crashing against the cliffs along the coast. As the ocean pounded the small boat, Bunk and DeLano were the only members of the hunting party to avoid becoming seasick.

Hoonah was a quaint fishing village inhabited by Tlingit, an indigenous Alaskan people. Bunk found their customs and beliefs both interesting and unusual. He wrote in his field notes about the local cemetery with above-ground "death-houses" that contained urns of water and food to aid the departed as they traveled on death's journey. There were also colorful totem poles with carved, painted figures representing religious deities. The Tlingit prided themselves on their handmade baskets, woven from thin lengths of bark stripped from the roots of young spruce trees. Bunk bought a lidded basket with a bold geometric design in warm hues of red, orange, yellow, and brown as a souvenir.

After the hunting party returned to Juneau, they transferred their gear onto another two-hundred-foot tramp steamer named the SS *Alameda*. When they departed for the Kenai Peninsula, the Inland Passage no longer protected the ship from the open ocean, where winter storms were boiling the sea. According to Bunk, for days on end fifty-foot waves tossed the steamship around like an "egg shell." Everyone in the hunting party except for Bunk, as well as a majority of the passengers, spent most of the time hunkered down below deck with seasickness.

In the week it took the ship to travel to the Kenai Peninsula, it stopped at a number of ports along the way to drop off and pick up passengers and freight. One stop was Cordova, Alaska, where Bunk and Plotts took advantage of some free time to ride the train thirty miles up the Copper River to view and film two gigantic glaciers. On the east, the Childs Glacier faces the much larger Miles Glacier on the west. Between them the river spreads out in a fifteen-mile-wide delta.

The railroad existed here only because J. P. Morgan and the Guggenheim family needed it to haul copper to the coast from their mine farther inland at Kennecott, and it would not be completed without building an extremely long bridge to cross the river and the delta. During the construction of the railway, the project engineers had serious concerns about how the glaciers would affect any bridge built in that location. Glaciologists had confirmed that at times over the previous two centuries the two glaciers had actually met and fused, something that would destroy the bridge if it were to happen again. As late as 1885 the nose of Miles Glacier was only about 120 feet from the proposed location of the bridge, but by 1908 the glaciers had retreated to what was considered a safe distance of about three miles. In early 1909 the investors had delayed building the bridge, waiting to see if the glacier was growing or retreating. However, when the government passed a law requiring any railway construction projects to be completed within four years or be subject to onerous fines, construction of the bridge began at a breakneck pace.

To traverse the river at that specific location, it was necessary to design and build a four-span suspension bridge. Such bridges required the building of massive caissons, which were sunk into the river. After the project's completion, these structures just looked like supports, but during construction they housed workers who were digging down to bedrock below the icy water level. Not only was that a difficult and unpleasant job but the weather conditions near the glaciers were unbearable. According to a National Forest Service sign on the bridge today, "workers braved fierce winter weather. During construction, snow accumulated thirty-four feet, temperatures plummeted to −60 degrees F, and the winds howled up to 95 mph. Just imagine . . . seeing the spit from your tobacco 'chaw' freeze before it hit the ground, or feeling your way across a catwalk in gales as furious as a hurricane." Because of the complexity of the project and the rush to completion, it ended up costing $1.4 million to build; hence it was nicknamed the "million-dollar bridge."

After the bridge was completed in midsummer of 1910, the worries about the encroaching glaciers continued. Childs Glacier was growing eight feet per day and was only a quarter of a mile away from the bridge. The owners of the railway and mining company watched nervously as the glacier continued to advance on their investment. Finally, in June 1911 it was announced that the glaciers had begun to retreat. The glaciers never again threatened the bridge during the time it was used to haul copper for the next three decades. Although the investors spent $1.4 million on the bridge for the railway in the middle of nowhere, it was reported that they had harvested over $200 million worth of copper by the time the mine ceased operation in the late 1930s, when the price of copper tumbled.

Bunk described Miles Glacier as over 250 feet high and three miles wide at its face when he and Plotts arrived at the bridge. They saw slabs of ice slide off the face of the glacier, sending waves rolling across the river several feet high. Later that day they rode the train back to Cordova and the *Alameda* to resume their journey to the Kenai Peninsula.

Several days later, when the SS *Alameda* finally entered Cook's Inlet, they still had a ways to go. Cook's Inlet runs along the west side of Kenai Peninsula and is about two hundred miles in length and ten to thirty miles across. The destination for the hunting party was a salmon cannery approximately halfway up the bay on the east side, at the mouth of the Kasilof River. There they planned to meet their hunting guide and a group of packers who would lead them fifteen miles up the river to the glacier-fed Tustumena Lake.

Upon arrival at the mouth of the Kasilof River, they were surprised to see a towering three-masted ship looming ahead of them, anchored just offshore by the cannery. The ship, named the *Star of Russia*, was owned and operated by the Alaska Packers Association (APA), a trade organization established in the late 1800s by the salmon fishing industry to control the supply of canned salmon so they could raise the prices. Eventually the APA became so successful that at one point they generated 80 percent of Alaska's revenues, thus making them exceptionally influential in the politics of both the territory of Alaska and the United States.

In 1908 the APA purchased the *Star of Russia*, their very first iron-hulled sailing ship. They preferred sailing those tall ships because the operational costs were much less than those of steamships. The association liked the name so much that they named many of their other ships following the "Star of" pattern: *Star of Alaska* and *Star of Finland*, among others. At the beginning of each summer, these tall ships would ferry the crews from San Francisco to the various canneries along the coast of Alaska. After the ships were filled with canned salmon, they would return to San Francisco and unload. At the end of the season, about the time when the hunting party arrived, the massive ships would load up the final production of canned salmon, collect the crew, and return them all to the "lower forty-eight."

The cannery was a small village, the proverbial company town. It had everything the worker could want or need: the canning facility, housing for the workers, a small hospital, and a company store. Being extremely cost conscious, cannery companies utilized cheap labor, mainly migrant workers from China, Japan, and the Philippines. Despite its image as a price-fixing monopoly, the APA displayed benevolence in the spring of 1919, when, during the worldwide Spanish flu epidemic, it opened many of its cannery hospitals to the native population in Alaska, buried the dead, and provided for orphans of those victims of the flu.

The SS *Alameda* was too large to dock at the cannery, so a salmon tender was dispatched to bring the hunting party ashore. The superintendent of the cannery welcomed the new arrivals with open arms. He immediately informed the hunters that their guide and his native Alaskan helpers would be delayed. They were coming by small boat from Kenai, a small village fifteen miles to the north, and rough seas in Cook's Inlet had postponed their departure. In the meantime, the men were invited to move into the cannery. The three doctors would stay in the hospital and the rest of the hunting party settled into the superintendent's quarters. After supper that evening, the hunting party gathered around a campfire and sang songs and told stories, led by the doctors and DeLano.

The next morning, there was still no sign of the guide and his helpers, so the hunting party explored the cannery village and the surrounding area. Bunk wrote about bartering with a local resident for a bear skull and DeLano wrote about the local insects. He quipped "that the main difference between the Alaskan mosquitoes and others is that the former have a white spot between their eyes about the size of your hand." DeLano also bemoaned the notorious "no-see-ums," a nearly invisible insect known for its stinging bites. He reported that in the early days, before there were jails, criminals would be put in a mosquito/no-see-um-proof box without any clothes — there was no chance they would try to escape for fear of being attacked by those pesky insects.

On the morning of August 24 the hunters awoke to find that their guide and six packers had traveled all night in open sea from Kenai and arrived before first light. Emil Berg was the leader of the guide party, and an accomplished and experienced guide. He was the younger brother of the famous Alaskan guide Andrew Berg, who first explored and hunted the Kenai Peninsula, including the Tustumena Lake. During the late nineteenth century, moose antlers and other wildlife trophies became popular in the "lower forty-eight," causing an influx of hunting parties in need of guide services. In 1908 the federal government established game limits and a system to license guides for the Alaska Territory. Andrew was issued the first guide license, "No. 1," in 1912. His younger brother, Emil, was licensed in 1914.

The statute specified a pecking order for guides and stipulated two classes: first-class guides were white citizens of the United States and second-class guides were "men of mixed blood leading a civilized life, Indians, Eskimos or Aleuts, all herein referred to as natives." The law also set the rate that could be charged for the guide service: no less than $5 per day and no more than $10 per day for both first- and second-class guides. In addition, the statute recognized a category of packers, who would do much of the heavy work such as carrying supplies, rowing boats, cleaning carcasses, and cooking. Packers could be paid no more than $3.50 per day.

Realizing that the entire hunting party of thirteen could not travel in one dory, they borrowed a second boat from the cannery superintendent. As they began to pack, the guide told them that their departure from the cannery was going to be delayed because the tide was running out of the river. All marine activity in Cook's Inlet depended on the tide. Due to the geography of the inlet, high tide rises forty-five feet above low tide, the second-highest difference in the world. That tide differential causes the water currents of the rivers pouring into the inlet to be extremely strong, so they were forced to wait for the tide to come in to push the current and the boats up the river. This delay was just as well, because it allowed them to better assess their supplies and equipment. Although the guide brought what DeLano described as $500 worth of food and provisions, the party decided they needed to buy additional slabs of bacon and three hundred pounds of flour from the cannery superintendent. Many of them also wisely purchased rubber boots and hip waders when they learned how they would travel up the river.

CHAPTER THIRTEEN

Up the River to Sheep Country

In the early afternoon when the tide began to come in, some of the packers tied the two boats together and put an outboard Evinrude motor on the lead dory. When everything was finally loaded, the hunting party pushed off and headed up the river, which was over two hundred feet wide. With the aid of the boat motor and the tide gently pushing them upstream, the rowers expended little energy for the first three miles. When they reached the rapids, however, the current was too strong to continue with the motor and oars. From that point on the hunters and their guide party would have to resort to what was called "lining up" the river. Both boats were untied and a hundred-foot length of rope was attached to the front of each boat. One man stayed in each boat serving as a steersman, while the hunters and packers grabbed the rope and pulled each boat upstream from the shoreline. According to DeLano, each boat filled with their equipment must have weighed three thousand pounds.

Under the best of circumstances, the men would have strained to pull the heavy boats, but the existing conditions for lining up the river were treacherous. Not only was the footing along the bank of the river slippery but the hunters and packers constantly had to navigate around bushes, through ditches, and over fallen logs and rocks. Whenever trees presented themselves at the edge of the river, they had to maneuver the ropes around them. After just a few miles, the men began to feel the pain of sore muscles from the heavy pulling and cuts and bruises from frequent falls.

Later that afternoon rain began to pour, so they pulled into a cove to establish camp for the night. They set up tents and sleeping tarps and started a fire to prepare dinner and take the chill out of the air. After a hearty dinner of bacon, coffee, potatoes, bread, corned beef, and canned fruit, they sang and told stories, as usual led by the doctors and DeLano.

A steady cold rain continued the second day as they broke camp and resumed lining up the river. In that stretch of the river, trees and brush fre-

Lining up the Kasilof River in Alaska, 1919 (Alaska State Library, Dr. John Outland, Alaska Big Game Hunt, 1919 Photo Collection, P425-17-40)

quently made the banks impassable, forcing the pullers to wade into the water, to either pull or carry the end of the rope around an obstacle in the riverbank. Worse yet, sometimes they either ran out of land or needed to avoid faster and stronger currents, forcing them to carry the rope to the other side of the river. Every time they had to change sides, it was demoralizing because it involved being pushed back downstream, losing some of the hard-fought ground they had previously achieved.

Bunk, Plotts, and Teal did not have waders, so they walked ahead along the bank to set up the campsite for the noon meal. The walkers found a cove with an open place to set a fire for dinner. When the boats had not arrived as soon as expected, Bunk walked back down the river and soon saw the first boat just coming around a bend. He shouted to the men in the boat that a campsite had been located and their boat should reach it by noon. The men replied that the second dory was about a half a mile behind and that Dr. Outland had fallen into the current. They said that if they had not cut the rope and thrown him a line, he might not have survived.

After lunch, everyone joined in the pulling. After a while Bunk gave it up because he worried that the still camera he was carrying might get wet. Also, blisters had formed on his heels from his new boots. As it continued to rain, Bunk walked on ahead. With evening approaching, Bunk looked downstream and saw that because the boats had changed sides of the river several times, he was now on the opposite bank from the boats and the men. He knew it was too dangerous to spend the night on the wrong side of the river, because he had seen numerous bear droppings in the area, so he needed to devise a plan to get to the other side. As he stood there shivering, soaked from the constant cold rain, he looked a short distance upstream and saw a hunter, not from their party, who was camped on the other side.

As he pondered his dilemma, the first boat came up the river and pulled in on the stranger's side of the river. With that boat was their guide, Emil Berg, who had noticed Bunk stranded across the river. Berg borrowed the hunter's empty boat and poled his way up the stream along the back current near the shore. When he reached a hundred yards, he turned the small boat downstream and shot across the river, losing only fifty yards. Bunk ran upstream and jumped into the boat, which shot back across the river, landing at the exact spot where Berg had started. Bunk was so impressed with Berg's ability to navigate the swift current that he wrote in his field notes, "Only one man in a thousand could have done such a masterful job."

After the second boat arrived, they erected their tents, built a large bonfire, and started supper next to the camp of the stranger. After supper, in spite of the tiresome day, they somehow conjured up the energy to sing songs before going to bed. As they bedded down for the night, it was still raining, and some slept wearing wet clothes. The stranger who shared his campsite with the hunting party was heading downstream, so he was the last human contact they would make until they returned to the cannery.

On the afternoon of the third day of lining up the river, the current subsided when the boats reached the juncture of where the river joined the lake. At last their lining-up battles were over. Berg had the packers cut a small tree to make a mast and rigged a tarp as a sail. The two dories were tied together and with wind, motor, and oars, they headed up the river for the last four miles before entering twenty-five-mile-long Tustumena Lake. The hunting party found the lake surrounded by a majestic wilderness of flat plains rising slowly up to the mountains on either side about ten miles away. Much of the shoreline was swampy and home to an abundance of moose and bear. At the far end of the lake, the gigantic Tustumena Glacier was surrounded by an expansive mountain range with five-thousand-foot peaks and bighorn sheep.

After sailing, motoring, and paddling for fifteen miles on the lake, the party arrived at a log cabin in Devil Bay around dusk. As they prepared to spend the night, Berg explained that this log cabin was one of several around the lake built by his brother. “Hunter’s cabins,” as he called them, served as a refuge for hunters and trappers. Even though the cabins were vacant, they were well stocked with flour, matches, a tin stove, and dry kindling. He told them that the etiquette for travelers was to leave it as you found it. No one would ever consider violating that rule because such cabins were lifesavers for travelers and hunters, especially if they got caught in an unexpected winter storm on the Kenai Peninsula.

The following morning, the party reloaded the boats and resumed their journey toward the glacier. After a short distance, the Evinrude motor sputtered and died. After repeated tries to get the motor started, they gave up and continued on by rowing and using the sail for the remainder of the morning. At one o’clock they finally arrived at their first permanent campsite, a second cabin located near the mouth of Crystal Creek and about a mile short of the glacier.

Once camp was established, they spent the rest of the day exploring the area around the campsite. Crystal Creek was only seventy-five feet from the camp, so Outland, Teal, DeLano, and Masson took their fishing rods upstream to try their luck. Upon their return, they announced that the stream was filled with bright red salmon, some alive, but many dead. For freshwater fishermen from the Midwest, that sight was unexpected. They learned from the guide that after salmon are born in small streams like Crystal Creek, they swim downstream to the ocean. Four years later they return to the small stream where they were born. By late summer they change color to protect them from ultraviolet light while swimming in the shallow waters of these spawning streams. After they reproduce, the salmon die.

Later that afternoon one of the packers named Mike was complaining of a stomachache. Dr. Masson examined him and suggested he might have appendicitis — or it might just be gas. Bunk wrote that night about his concern that there might be "an operation as a little excitement." At bedtime Bunk offered Mike some cinnamon gingersnap cookies. The hunters slept restlessly that night in anticipation of stalking sheep the next day.

A steady rain greeted the hunters when they awoke early the next morning on August 28. Although it looked like the rain would continue all day, their spirits were high in anticipation of their first day of the actual hunt. Berg asked for a volunteer to go with him to explore where the sheep were ranging. Everyone wanted to go, so they drew straws and DeLano won. Berg and DeLano grabbed their guns and immediately set off up the mountain. The others, along with the packers, were to load up what they would need to establish a temporary sheep camp and pack it ten miles up the trail toward the mountain.

Bunk was chosen to stay at the cabin, which suited him fine since his heels were still blistered and walking was difficult. Bunk's role on the trip was not to hunt but to serve as the custodian of all the animals they shot. Bunk's suspicion that the hunters were more interested in the excitement and adventure of big game hunting than in adding specimens to the museum's collection would soon be confirmed. On more than one occasion that difference in purpose led to friction between Bunk and the hunting party. The first example of that tension was found in DeLano's notes, when he summarized in a condescending manner that Bunk was the "undertaker of all game-to-be." After the hunters and packers departed to set up camp in the mountain, Bunk decided to clean some skulls, so he searched for a large pot in which to boil them. Since none were found in the cabin, he boated over to another cabin across the bay and found a suitable kettle.

As Berg and DeLano took off up the mountain to scout for sheep, the march proved to be more physically demanding than DeLano had anticipated. Berg was a seasoned walker and in much better shape than the big-city attorney. When the guide kept pushing the pace as they climbed over rocks and brush, higher and higher into the mountains, somehow DeLano managed to match Berg's pace. Later he claimed that his ability to keep pace was due to the "hardening" he got from lining up the river for three days.

Within the first hour they came across a black bear foraging on the vegetation on the side of a nearby hill. The bear didn't see the hunters, so they moved slowly and quietly around the hill to get closer, but always making sure to stay downwind. After half an hour of repositioning, DeLano both lost his bearings and became separated from Berg. Suddenly, as he crept around a boulder, he came face-to-face with the bear. Less than sixty feet away, the startled bear reared up on his hind legs, roared angrily, and charged directly at DeLano. All of DeLano's senses were on overload as they stared directly at each other. There was no doubt that the bear would get him if he didn't shoot it first. DeLano raised his rifle, and his first shot was successful. The bear dropped in a heap. Just as the first bear went down, a second bear popped up from behind a nearby rock. Although the second bear was close and looked as if it were going to attack, DeLano was still recovering from the adrenaline rush from shooting the first bear. Officially, Alaskan guides are prohibited from shooting any game animals when conducting a guided trip, the presence of imminent danger being the one exception. Berg concluded that DeLano was not prepared to get off another shot, so the guide quickly raised his rifle, took aim, and fired, instantly killing the second bear.

DeLano was still reeling from the excitement of his encounter with the two bears when two young cubs scurried into view. The cubs posed no threat to the two men, so they let them go. To big game hunters, a cute little bear cub was not appropriate for a hunting trophy. Later they would learn that the decision to let the bear cubs live would irritate Bunk to no end.

As DeLano and Berg were skinning the two adult bears, Masson and two of the packers made their way up the trail. After DeLano explained the excitement surrounding the killing of the two bears, the guide looked at his watch and asked if they wanted to keep going. DeLano was tired and ready to return to camp, but wrote in his notes that he erred by pausing after saying, "Sure," thus giving the guide the opening to jump up and take off before DeLano could finish his sentence. According to DeLano, what he had tried to say was, "Surely it is time to beat it for camp." But with the loss of the opportunity to call it a day, DeLano and Masson scrambled off at a fast pace following Berg to look for sheep, leaving the two dead bears in the care of the packers.

At the same time, back at the cabin by the lake, Bunk was busily taking care of what the hunters would have considered mundane chores. When it looked like rain, he went around the campsite gathering up drying clothes.

Later he used cyanide to kill several insects that he had collected for the entomology collection. When the rain started, Bunk retreated into the cabin and with nothing else to do, began to look through the food supplies. He was impressed that the guide and packers had brought such a large assortment of food, mostly by the case, and described what he found in his field notes. There was "corned beef, bacon, flap jack flour, oatmeal, condensed milk, coffee, tea, cocoa, tapioca, rice, raisins, catsup, Worcestershire sauce, cakes, prunes, potatoes, canned sweet potatoes, peaches, pears, plums, pineapples, and maple syrup." Bunk made no comment on how the Worcestershire sauce might seem out of place in all those provisions because he knew they would be eating a lot of wild game and it would help cover up any strong flavors they would encounter. Bunk wrote that it was typical for the packers to do the cooking and with Bill as the head cook they all ate "better than they did on the passenger ship *Alameda*."

Later that afternoon, two of the packers, Bill and Mike, returned to the cabin with the skins, heads, and feet of the two black bears that DeLano and Berg had shot that morning. After the two packers headed back up the mountain, Bunk began to work on the bears. He started skinning the head and feet of the young male bear, which he hung up outdoors to dry. Later, when Mike and Bill returned to camp, Bill announced that he felt sick with a fever and cold symptoms.

Back on the mountain, after miles and miles of hard hiking in the rain, DeLano, Masson, and Berg reached the top of the divide on the Kenai Peninsula and stopped briefly to take in the extraordinary view. When DeLano described the panorama in his notes, his tendency for hyperbole came shining through:

> As we started, the sun peered out from behind the clouds for a few minutes, and I wish I had the words at command to describe the most beautiful sight that I have ever seen. Miles away across the inlet [Cook's] giant [Mount] Iliamna, always snow-covered, poking his nose through a rift of the clouds, to scent the trail of the sun nearby; close at hand, the many-colored green on moss and shrub, gradually being feathered with a down of white, and so joining hands with the nearby peaks, already frosted from last night's snow; miles behind us. Tustumena Glacier sparkly with mirth at seeing the sun; miles in front, on the other slope of the mountains, a little river first climbing down step by step the range on range of frosty hills, then gliding into the lowlands, where it wooed the virgin green of the

> spruce and swamp and tundra, and so at last found a home in Kachemak Bay, which lay there calm and quiet as if it were doomed to be the final resting place of the rough scenes around; and all this seen prism-like through scattered, glistening snowflakes, was a sight long to be remembered.

When Bunk read what DeLano had written, Bunk said it was a good thing DeLano did not have "his words at command" or else his notes would have been a book.

For the remainder of the afternoon, the trio continued looking for sheep across several mountain ranges. They were disappointed to find only a few herds that were all ewes, which were protected from hunting. It was getting late, so they headed to the sheep camp for the night. They stumbled down the mountain four hours later, arriving at the sheep camp at nine o'clock. Back at the cabin, Bunk and Mike were left with the responsibility of looking after Bill that night because all the doctors were ten miles away up the mountain at the sheep camp.

The second full day of the sheep hunt brought success. Teal was experiencing knee pain, which Teal attributed to his "over-exertion in youthful prayer," so he and one of the packers went on an easier route and brought back one ram. Outland, Masson, and DeLano went off with Berg and found a herd of seven ewes and five rams. Outland held back, saying that he would not shoot until Masson and DeLano had gotten their ram. Once Masson had shot two and DeLano shot one, Outland and Berg left them to skin their rams, hiking further up the mountain to find a ram for Outland. After they finished skinning the rams, Masson and DeLano covered the carcasses with rocks. Gathering up their ram heads and skins, they hiked to the spot where Outland and the guide agreed to meet them. When DeLano and Masson arrived at the rendezvous, the weather had turned cold and snow began to fall in earnest. The two were not dressed warmly enough and had no idea how to get back to the sheep camp. They became concerned they would get stuck there on the mountain. Finally, as evening approached, Outland and Berg arrived. Their delay was due to Outland having killed the largest and finest ram of the trip.

While the hunters were enjoying their success up on the snowy mountain, it was raining down at the cabin, and Bunk was worried because Bill was showing no signs of improving. He wrote, "I could be happy alone in this cabin on a beautiful lake with mountains behind me. How often I have wished my self [*sic*] in such a place and now I have an Indian with what I think is pneumonia to take away the charm." When two of the packers came down the mountain for more supplies, Bunk sent a note up to Doc Outland to say that he could use some medical help with Bill. Bunk hoped one of the doctors would come down that evening, because, as he put it, he did not want Bill to "die on his watch."

In the meantime, Bunk salted the bear skins and continued to boil their heads. In his field notes, Bunk commented that his feet were considerably improved: "My heels have not hurt me quite as bad today and I hope they are on the mend." As he worked, he watched the birds around the campsite, mostly of the larger variety: magpies, gulls, one bald eagle, hawks, and some mallard ducks. There were also a few smaller birds flying around the camp. Bunk wished he had his shotgun with the auxiliary barrel, so he could add some specimens to the collection. But there was no room for it on this trip. Furthermore, it was clear that "the guide and the rest of the party do not see the value of it." Only the large game was important to them.

By evening, with no sign of the doctors, Bunk grew irritated. He concluded that the doctors had little faith in Bunk's concern and thought Bill was feigning illness, or that the doctors were more interested in hunting than in the well-being of the Alaskan. That evening he tended to the ailing man as best as he could by rubbing Bill's chest with a mentholated ointment, placing a rag soaked with turpentine on his chest, and giving him a slow-acting iron pill and aspirin. That night Bunk fumed, "I hope he will not die til morning as I do not fancy playing undertaker. . . . I don't believe anyone will be down tonight although if I had got the letter I sent to Doc with Mike this morning I would have come down through fire, but not so with this bunch of crazy sports."

On the following day Outland finally came down to the cabin to check on Bill. The rest of the party was still trying to make sure everyone had a ram, and would not be returning from the sheep camp until the following day. Seeing that Bill was improving, Outland went to bed, as he was all done in from two hard days of marching around the mountains hunting sheep.

Later that day, when Outland told Bunk and the packers about DeLano's adventures of shooting the two black bears, Bunk learned that they had passed on shooting the two cubs because they felt sorry for them. From Bunk's perspective, adding an entire family of bears to the collection would have been an extraordinary coup for the museum, and passing up the opportunity made him furious. Bunk said in his notes, "I am clear out of luck in this party. I am not supposed to hunt and the others just get what [they want] for themselves. If things don't change, I will see them in —— when it comes to getting me to fix up their [trophy] heads." To the doctors and DeLano, big game hunting was for the thrill and the sport, while Bunk wanted the trip to be about science and increasing the collection at the natural history museum. And that difference of opinion would continue to undermine the rest of the trip. When DeLano learned of Bunk's displeasure with them for not taking the two bear cubs, DeLano dismissed Bunk's concerns with the following: "We found that our tenderheartedness was entirely wasted upon our taxidermist."

CHAPTER FOURTEEN

Moose and Bear Country

On day four at Crystal Creek, the hunting party came down the mountain early to the cabin by the lake. All joined together to quickly pack the boats so they could move to moose country on the other side of the lake. After casting off, they had to row, since the outboard motor still wouldn't start. Twelve miles across the lake they found another rescue cabin in among the spruce just fifty feet from the shore of the lake. Bill was not fully recovered, so Bunk took the time to get him settled in and comfortable. Bunk felt Bill would be good as new in a few more days.

The next morning, Emil Berg, John Outland, and Raymond Teal went to scout the area for moose and grizzly bear. At two o'clock, Outland and Berg came back to camp and proudly reported that Teal had wounded a large moose and they wanted Plotts to come and shoot a movie while it was still alive. When asked how far the moose was, Doc Outland joked that it was about twice as far as the bear. Outland explained that on the way back to camp a "thousand pound" grizzly bear had "gotten in the way of his bullet" and, according to DeLano, they needed "coroner Bunker to come and hold an autopsy."

According to Bunk, with that announcement the camp began to boil with activity. "I got my skinning tools, forgot my [sore] feet and the Indians got their packs and soon all of us . . . were on the move." After a three-mile hike "over fallen timber and holes and bumps, swamps and every kind of ground," they reached the bear. It was a huge grizzly, weighing over a thousand pounds. Bunk and two of the packers immediately began to skin the bear while the rest of the party went on to the moose. Because of the bear's massive weight, the three men could skin it only by rolling the bear from side to side. After working for two hours, Bunk and the men packed up the skin, skull, and leg bones from one side, and walked back to camp, arriving shortly after six o'clock. The party that went off to retrieve the moose did not return to camp until after nine o'clock. They left the moose, as well as the motion picture camera, because it got too dark before they were done, and they were too tired to carry anything back to camp.

That evening another packer complained of feeling sick to his stomach. Once Outland and Masson concluded that the packer was faking his illness, they pretended to diagnose his problem as appendicitis and threatened surgery. Teal taunted the surgeons, saying that they always wanted to "cut a person from hell to breakfast." When Bunk naively offered some standard remedies for stomach pain, DeLano reported, "Lay Brother Bunker came forward with his never-failing, all-curing, patent home remedies, at which all others howled [in] derision." Bunk took the kidding in stride, but he was growing weary of the jokes at other people's expense. Right after Bunk decided to call it a night and went to bed, the packer with the stomachache quietly left camp for the night. When he returned the next morning, there was no mention of his pain or illness.

After breakfast, DeLano, Masson, and Outland went bear hunting with Berg. They had lost interest in the moose from the previous day, so Plotts and several packers set off to bring back the moose and the movie camera. After a short while, Outland came walking back into camp. He explained that his gun had accidently discharged on its own, so Teal and Bunk helped him fix it. Afterward Bunk returned to working on the bear head and feet for the remainder of the morning. After dinner, he worked on the moose that the cameraman and the packers brought back to camp. That evening before supper, the hunters returned to camp and announced that they had killed another grizzly bear, but left it to be brought back later.

When they awoke the next day, the weather was changing. The wind had come up during the previous night, bringing the waves of the lake to just ten feet from the tent, and it looked like it was going to rain. Not to be discouraged by the weather, Outland, DeLano, and Berg got an early start and went off to hunt. Bunk took advantage of the quiet camp to wash the hide of the large bull moose. He tied a rope to the hide and put it out in the waves of the lake. When he went to pull it back in, it proved to be an exhausting struggle because the hide weighed about 250 pounds without any water.

For the three hunters, the day would be both successful and exciting. First Outland shot another fine bull moose, but this one had horns with a sixty-three-inch spread. Next, they spotted a bear on a distant hill. All three took shots at the bear, which seemed to have no effect on him other than making him turn and charge toward them. Suddenly, the bear disappeared from view in a valley between them. The three hunters started running in

Bunk skinning a moose (Alaska State Library, Dr. John Outland, Alaska Big Game Hunt, 1919 Photo Collection, P425-17-54)

the direction of the bear when, without warning, the bear reemerged, coming at full speed over the hill, right toward them and just thirty yards away. With their adrenaline surging, all three fired just in time to drop the bear before he reached them. After bagging a moose and having a close call with a bear, the hunters went back to camp with great satisfaction to enjoy a supper of ribs and tenderloin of moose.

The next morning was scheduled to be the last day of hunting, since they were starting back to Kasilof in two days. Bunk cautioned the hunters about needing a down day, one where nothing was to be killed, to give him time to prepare the hides so they would not spoil during shipment. But such advice fell on deaf ears with the sportsmen. Right after breakfast, DeLano, Masson, Berg, and three packers went out one more time for moose. Bunk grumbled about how a new kill would make preparing the skins more difficult, while Outland and Plotts stayed in camp and played cards. The hunting party located a cow moose about 475 yards away from camp and successfully fired a killing shot. As they moved closer to the dead moose, they saw and killed a calf, thus completing the hunting, or so Bunk thought.

September 5 was to be a day off for everyone except Bunk. The hunters and packers were all exhausted from day after day of the grueling hunting routine: up every morning, rain or shine, walking miles before finding game to shoot. Hauling enormous amounts of equipment and animal parts took considerable energy, as did those travel days of turning an oar and moving tent and equipment to remake a camp. As they prepared for returning to civilization, Bunk was busy trying to finish up all the spoils of their hunting. He got most of his help from the packers, since the hunters were only interested in the supply side of big game hunting. Bunk lamented about having to leave the bones of three animals in the timber because it was more than he could clean before their departure. For the hunters it was all about the adventure of the kill; they were not concerned whether the "kill to clean" ratio was even in the end.

The next day they planned to move their dories twelve miles closer to the headwaters of the Kasilof River and set up camp, poising themselves to descend the river back to civilization. But Teal and Masson wanted one more chance at getting a bear, so they took Berg and one of the packers and walked overland toward that final campsite. The rest of the crew set off with the two boats. At a point about halfway from the camping destination for that night, they left one of the boats for the hunters. As for the hunters, along the way they found a large black bear feasting on the carcass of a moose they had previously killed. They shot and skinned the bear and loaded it into the boat that had been left for them. That night they joined the rest of the party, giving Bunk one more trophy to clean.

A cold steady rain escorted them as they rowed from the lake to the head of the Kasilof River. As the current began to flow faster, they pulled their dories into a cove, untied the two boats, took down the sail, and discarded their makeshift mast. After they made certain that their cargo was secure, they rowed back into the fast-moving current and held on tightly for an exciting ride down the rapids. With great skill the experienced packers safely steered the small boats through the boiling current and around the rocks. What took three gut-wrenching days going upriver only took hours to descend. It would have taken two hours to complete their fifteen-mile journey down the river if it had not been for Outland insisting that they stop to take photographs and movies that re-created their lining up the river and shooting the rapids.

On September 7 they arrived back at the cannery to see that the *Star of Russia* had departed for San Francisco. The cannery was eerily deserted except for a lone watchman who would look after the little village all winter. There they settled up with Berg and the Native Alaskan packers. To return to the lower forty-eight, plans had been made for a small boat to carry the hunting party one hundred miles up the Cook Inlet to Anchorage, where they would catch a steamship on September 12. However, on September 9 they were informed of the boat's delay due to another storm hitting the inlet. Bunk worried that they might miss their ship in Anchorage. Finally, on September 10, their boat arrived at about one thirty in the afternoon with a small boat in tow. By two o'clock they were loaded up and heading to Anchorage. The boat had plenty of power with its twin four-cycle engines, but its flat bottom allowed it to roll with the waves. Bunk described the voyage as "like [riding in] a washing machine." Once again everyone in the hunting party, with the exception of Bunk and DeLano, became seasick. Finally, after a harrowing sixteen-hour ordeal, they arrived in Anchorage at six o'clock the next morning, September 11, 1919.

At the docks of Anchorage, Bunk learned that the moose heads needed to be packed in a crate before they would be accepted for shipment. He took the heads uptown and had a wooden box built that ended up weighing 365 pounds. On September 12 the party learned that the steamship was late, so their concerns about missing the ship were for naught. The steamship finally arrived in Anchorage early on September 13. The crates of wild game began to be loaded at 7:00 a.m. and Bunk and the hunting party were on board by 12:30 p.m. Bunk kept his watchful eye on their crates of animals because much of the cargo was removed and reloaded several times. By the time the ship left the dock at 5:00 p.m., Bunk was satisfied that all his freight was on board.

A storm came up as they steamed out of Cook's Inlet, causing the ship to rock wildly. Because the storm continued through the next morning, most passengers missed breakfast. On the second day the rough seas were worse. Now the swells were even larger, with the bow of the ship rising and falling fifty to seventy feet. Finally, by midafternoon they reached the calm waters of Resurrection Bay and Seward, Alaska, where the ship docked for a few hours. Around suppertime the ship returned to the heavy rain and rough seas. This was the only night that Bunk got seasick during the entire trip. According to his notes, "7 pm the boat is just beginning to pitch a little and people are beginning to slip away to their rooms. I think for the first time I will do the same as I ate a big piece of watermelon for supper and it sowered [*sic*] my stomach right away and I don't feel as if I would stand much pitching."

As the sun approached the horizon late on September 19, the hunting party reached Ketchikan, Alaska, where Bunk cleared all of the specimens through customs without any problem. Two days later, on Sunday, September 21, their ship sailed into Seattle, where their adventure came to an end.

Over six weeks had passed since they had boarded the steamship bound for Juneau. Bunk's extraordinary adventure resulted in the addition of a variety of wild animals to the specimen collection at KU: three ermine, two otter, four porcupines, two mountain sheep, four black bears, and three large Kodiak bears. Most of the specimens added to the collection were listed as the skulls only; however, there were several moose skins added as well.

Once he returned to Lawrence, Bunk had time to reflect on the Alaska expedition. In hindsight, he would have preferred sharing the grandeur of Alaska with his old friend Theo rather than that group of exuberant sportsmen. Bunk also remembered how he and Theo never shared a harsh word, despite being joined at the hip back in 1911 and 1912. In contrast, Bunk frequently felt at odds with the hunters, especially when the doctors refused to abandon their own adventures up on the sheep mountain, leaving Bunk to care for the sick packer down at the base camp. Also, it annoyed Bunk that they never valued, or even pretended to appreciate, Bunk's perspective on hunting for the advancement of science, such as when DeLano and the guide only considered the two bear cubs as poor trophies.

Bunk had difficulty understanding how the doctors and DeLano looked at hunting. To a hunter, the act of hunting was a means, not an end. To them the hunt, with its skill and adrenaline, was of greater importance than the dead animal that merely served as proof of the hunter's conquest. According to José Ortega y Gasset, a noted Spanish philosopher and essayist, "one does not hunt in order to kill; on the contrary, one kills in order to have hunted. . . . If one were to present the sportsman with the death of the animal as a gift he would refuse it. What he is after is having to win it, to conquer the surly brute through his own effort and skill." Bunk, on the other hand, would gladly accept a dead animal, because adding a specimen to the museum collection is more important than the act of hunting.

In the end, despite Bunk's frustration with the hunters, a deal was a deal. As a result, the hunters ended up with bearskin rugs and mounted sheep heads. Years later Bunk realized that his reaction to their confident and thoughtless behavior most likely related to the recent loss of his friend. Bunk was usually easygoing—he didn't let things get under his skin. But when Theo died, Bunk suppressed all his emotions. The insensitive actions of the hunters merely struck a nerve with Bunk that allowed his emotions to bubble over. If he had had more time to mourn the loss of his friend, he might have not been so frustrated.

CHAPTER FIFTEEN

Bunk's Boys, Campus Politics, and Beneficial Beetles

The American economy flourished in post–World War I years, at least in the populated cities. Unfortunately, Kansas and other agricultural states suffered after the war due to the drastic drop in grain prices. When the fighting in Europe interfered with European grain production, US farmers ramped up their grain output to meet the demand. However, once the war ended, grain production resumed in Europe, causing the worldwide supply to outstrip demand, which resulted in the collapse of grain prices. A poor agricultural economy translated in the continuing fiscal pressure on midwestern universities like KU, which trickled down to further stifle Bunk's meager paycheck.

Bunk turned fifty years old and was in the prime of his career at the museum as the 1920s began. On the home front, Fedalma was now a young woman, an unabashed cat lover who dabbled in assorted courses at the University of Kansas. As a headstrong young woman, she refused to take any other courses besides those that interested her. With only classes in art, literature, and philosophy, she made little progress toward a degree. Her social life was limited to the top floor of Strong Hall, with the art students and faculty. As an elementary school student, Audrey ran in the circle of several popular girls who lived in the south part of Lawrence. Her close friends were the daughters of professors and the owner of the local newspaper. Audrey was more social than her sister, but took after her father in that she was more comfortable as a follower than a leader in her group of friends. Bunk and Clara fell into regular routines that revolved around Bunk's work schedule: weekdays in the museum workshop, weekends hunting at the farm, and those extended collecting trips during the summers. Although Bunk was close with his wife and girls, he still spent the majority of his time working with the university students, making sure they learned how to increase the collection of birds and mammals.

In March of 1921 the University of Kansas published its *Annual Catalog*, which suddenly declared Bunk a "man of letters." Every year the university produced the catalog to impress the board of administration, and in turn the state legislature, as evidence that the university is being well managed. The board, which by this time had replaced the board of regents and possessed a wider range of administrative responsibilities than just the state universities, consisted of three members, all appointed by the governor as part of a notorious patronage system. The university administration considered the board exceedingly political, stingy, and vindictive. The supplement to the *Annual Catalog*, titled "General Information," listed Bunk as an "assistant instructor" even though he had no classroom responsibilities. However, the real puffery involved showing that Bunk possessed a bachelor of arts (1901) and master of arts (1906), both from KU. Not only were those degrees fictitious, Bunk had been living and working in Oklahoma in 1901. A few years later, the information about the degrees was eliminated from subsequent annual reports, though Bunk always joked that he was tempted to ask for a copy of those diplomas.

During the early part of the decade it seemed like Bunk was running a recruiting program for the University of California, Berkeley. Four outstanding students, whose time at the University of Kansas overlapped, went on to attend graduate school at Berkeley after earning their undergraduate degrees at KU. Berkeley was home to the well-respected Museum of Vertebrate Zoology (MVZ), founded in 1908 by the noted field biologist and zoologist Joseph Grinnell. The first of those KU students headed to California was E. Raymond Hall, who entered Kansas University as a freshman in the fall of 1920. A native Kansan, Hall, as Bunk always addressed him, was born and raised in Imes, a small town forty miles south of Lawrence. Despite being short-statured, Hall possessed a persuasive personality that convinced Bunk that one day Hall would be a leader in the field of natural sciences. During his time at KU, Hall earned a reputation for his critical thinking, dedication as a field collector, and thorough and copious field notes. After Hall graduated from Kansas University in 1924, he married Mary Frances Harkey, a fellow KU classmate, before moving to California. According to those close to Hall, he picked Mary as the "most attractive young lady from the Phi Beta Kappas in the senior class," thinking that between them they would have very smart children.

Next to venture from KU to Berkeley was a young man named Jean M. Linsdale. Growing up in Wellsville, Kansas, just south of Lawrence, he quickly developed a reputation as a prodigious hunter and collector. During his time at KU he collected over nine hundred birds and three hundred mammals. When Linsdale teamed up with Hall in the summer of 1921, they mounted a summer collecting trip to western Kansas, much like Bunk and Theo did in 1911. In 1925 Linsdale met Grinnell during his visit to KU, an opportunity that eventually led to Linsdale going to UC Berkeley. After Linsdale completed his doctorate, he joined the staff of the Museum of Vertebrate Zoology as assistant director and went on to be the resident director of the MVZ's Hastings Natural History Reserve, a 2,500-acre wilderness preserve and biological field station located in the Santa Lucia Mountains in Big Sur.

The third student who ventured to UC Berkeley was Ruben A. Stirton. He was born in 1901 in the small northeastern Kansas town of Muscotah, where he demonstrated a scholarly aptitude. When he enrolled at the University of Kansas, he studied zoology and participated in the wrestling team. "Stirt," as he was called, became a regular at Bunk's cabin. After earning his undergraduate degree from KU in 1925, Stirton was hired on as a mammalogist for two expeditions to El Salvador. After that field experience he followed Hall and Linsdale to continue his studies at Berkeley in 1927. However, by that time Stirton was interested in paleontology and studied under both Grinnell and the noted paleontologist William Diller Matthew. During those years he, like Hall, earned a reputation for meticulous field notes and catalog records, techniques he initially learned from Bunk. Ultimately, he earned his doctorate and went on to become the curator of fossil mammals at the University of California Museum of Paleontology in Berkeley. Eventually he was appointed as the fourth director of the museum of paleontology, a full professor, and chairman of the department of paleontology.

The final University of Kansas student with ties to Berkeley was another native Kansan named William H. Burt. He obtained his undergraduate and master's degrees at KU, then earned his doctorate from the University of California, Berkeley, in 1930. In 1935 he was hired by the University of Michigan as assistant curator of mammals and an instructor in zoology. Later he was promoted to curator in 1938 and professor of zoology in 1949, and gained a national reputation as a mammalogist.

All four of these students went on to distinguish themselves as naturalists and scientists and always attributed much of their success to the training they received from their old mentor, Bunk.

* * *

In any occupation, even one that someone really loves, the occurrence of unwelcome politics is always a possibility. Often it simmers under the surface until it erupts when least expected. One such event came to a head in the summer of 1921, when H. T. Martin marched into Bunk's workshop to inform him that the paleontology collection was coming under attack. Martin explained that Dr. Bennet M. Allen was up to his old tricks. Now both the departments of geology and zoology were trying to get control of the fossil collection. Although the conflict did not directly involve Bunk's collection (at least yet), he always cringed when territorial boundaries were challenged.

The fracas began when Allen asked the chancellor to approve his listing in the upcoming university catalog as "Professor of Zoology and Vertebrate Paleontology." Previously, Allen had been listed as merely a professor of zoology, but since he taught both courses, he felt entitled to the new designation. Allen also suggested that Professor Raymond C. Moore, who had previously been listed as a "Professor of Geology and Paleontology," should have "invertebrate" inserted before "paleontology," thus reflecting the course that he was teaching. Even though the letter may have seemed like nothing more than simple housekeeping, it set off a firestorm between the two professors.

The genesis of the feud began almost two years earlier, on September 13, 1919, shortly before Bunk returned from the Alaska hunting trip. On that date Chancellor Frank Strong announced his resignation, and from that point on KU's organizational structure began to shift. That news did little to alter the day-to-day activities of the majority of the university community, but for Bunk, who reported directly to Strong by virtue of the chancellor being the ex-officio director of the museum of natural history, that announcement meant he was getting a new boss. Bunk preferred the current administrative structure because Strong had always supported Bunk's control of the collection of birds and mammals. But more importantly, Strong's dual responsibility meant that the normal demands of university business occupied enough of the chancellor's time to give Bunk the freedom to manage the collection with considerable autonomy.

In May of the next year, the Kansas State Board of Administration announced the hiring of the seventh chancellor of the university, Ernest H. Lindley. Like his predecessor, Lindley lacked a background in the natural sciences — something that Bunk would have preferred. But it was his administrative approach that would ultimately change Bunk's world.

In the summer of 1921, when Moore got wind of Allen's proposed changes to his title, he immediately shot off a letter to the new chancellor complaining that without even being consulted, his title, and consequently his role with the university, had been diminished. Even though the new title accurately reflected what he was actually teaching, Moore felt rebuffed and wanted his academic title restored without the qualifier. In addition, now that the question had been raised about reorganizing paleontology, Moore offered his own power play. He boldly suggested that it made sense to expand his realm and have vertebrate paleontology join invertebrate paleontology under the control of his geology department. With that move, the two men were hopelessly deadlocked.

By 1921 the university bureaucracy had grown to the point that Chancellor Lindley felt it was time to delegate settling this dispute and established an ad hoc committee to resolve the matter. He asked the dean of the College of Liberal Arts and Sciences to chair the committee that would include the dean of the faculty and the two rivals, Allen and Moore. Both the zoology and geology departments presented written arguments to support why they should oversee all of paleontology. Although vertebrate paleontology had been closely aligned with the zoology department when Williston and McClung were in charge, Moore felt the academic discipline of geology was a better fit. Moore argued that his dominion was more appropriate because of the natural affinity between geology, the study of rocks, and paleontology, the study of fossils found in those rocks. In turn Allen argued that the reason for paleontology being part of zoology was more than merely historical. Such an alignment made sense because fossilized extinct animals are still animals, and hence should be under the jurisdiction of zoology. Finally, Allen suggested that putting paleontology in geology because fossils are a kind of rock was no more logical than having seals, whales, and marine birds under the control of the Bureau of Fishery because like fish, they all are found near or in the ocean.

When the dust settled, the ad hoc committee resolved the conflict by recommending that the interests of paleontology as a whole would be better served with "a union of the interests of Geology and Zoology than by conflict between them." Accordingly, Allen and Moore were given the new titles of professor of zoology and professor of geology, respectively, omitting any reference to paleontology. In addition, the geology department would continue to teach the invertebrate paleontology courses and the zoology department would teach the vertebrate paleontology curriculum. As for the overall administration of the collection, it was agreed that both department heads would be codirectors of the entire paleontology collection, and Martin would continue the direct supervision of the collection as assistant curator.

Martin was relieved that he remained in control of the fossil collection, but it is unlikely that either Allen or Moore must have felt any victory with such a King Solomon–like decision. Even if cooperation may be better than conflict, it does not happen just by wishing. Although the two rivals signed the committee's organizational recommendation to the chancellor, Allen felt rebuffed once again and left the university before the end of the school year.

Allen's departure left the zoology department without a chair, a position that Lindley needed to fill. For any savvy administrator, staff departures are always an opportunity to rethink and tinker with the organizational structure, especially with ever-present tight budgets. Lindley decided to look for a zoologist who could, in addition to lecturing and conducting research, handle the administrative duties necessary to reunite the various museum collections (birds and mammals, fossils, plants, and insects) under the leadership of one person. Furthermore, Lindley wanted someone with the academic credentials to give those collections standing as they related to the various scientific disciplines. By so doing, he could shed his role as director (albeit ex-officio) of the natural history museum. This would be a marked change for the administration of the museum, because from the founding of the university, the director of the KU Museum (or Cabinet) of Natural History had always been the chancellor.

In 1922 Lindley began a careful search for a new director. He found one candidate working at the University of Oklahoma, Dr. Henry Higgins Lane. Knowing that Bunk had contacts in Oklahoma, Lindley asked Bunk to check his references. Including Bunk in the vetting process was a wise move by the chancellor, since he hoped Bunk would be supportive of his new boss. On July 25, 1922, the Western Union office telephoned Bunk and read the telegram message from a former colleague in Norman, Oklahoma, with a good reference for Lane. Not too long afterward, Lindley hired Lane as head of the department of zoology and the director of the natural history museum. Although Bunk had previously enjoyed the idea that he reported directly to the chancellor, he had expected things to change eventually as the university faculty and staff increased in size. Besides, he was confident that Lane's dual role would occupy most of his time, so Bunk could continue operating the collection with independence.

* * *

Natural history collections rely a great deal on animal skeletons because they serve as the architectural infrastructure and support system of all animals, as well as evidence of the changes in their anatomy, functions, and mechanics during evolutionary time. For instance, when naturalists wondered whether plesiosaurs could have ever inhabited land as well as a marine environment, they reconstructed the vertebrae of the neck of the plesiosaur. Their findings showed that it was inflexible and projected forward. Given the length and weight of the head and neck, the plesiosaur could never have supported that part of its body without the buoyancy of water, thus proving that the species was aquatic.

Despite the importance of skeletons to natural history museums' collections of birds and mammals, the task of removing the flesh from the bones was tedious, time-consuming, and unpleasant. All throughout his career, like all museum preparators, Bunk had grown weary of boiling the skeletons and laboriously scraping them by hand. Such a manual process often resulted in damage to the more delicate skeletons. Moreover, the boiling action of the water often separated and rearranged the bones in the kettle, making accurate reassembly difficult. Bunk decided to take advantage of something occurring naturally in nature: insects eating the flesh of dead animals. Thinking that such a process would have a practical application in the museum, he had experimented for years with dermestid beetles, also known as larder beetles, to see if they could efficiently clean skeletons. Bunk began by placing the beetles and whatever bones he wanted cleaned in enclosed boxes. Eventually, through trial and error, he refined the process and settled on a procedure that resulted in a completely clean skeleton and required little human intervention. The entire process could take up to several weeks, depending on the size of the specimen, but by delegating the cleaning to the insects, collectors could now clean more than one skeleton at a time. Besides the advantage of the efficiency of the process, the beetles never damaged the bones, even for the smaller, more intricate skeletons.

After years of experimentation, Bunk finally wrote a three-page handwritten report on what he had learned about the dermestid beetle process. He never published the document and failed to put a date on it, making the exact date of when he perfected the process impossible to determine. However, it is likely to have been written sometime after 1921, because the museum collection of skeletal remains drastically increased in 1921 from 3–4 per year to an astounding 324, indicating the success of Bunk's new process. His document outlined the best practices when utilizing dermestid beetles in cleaning skeletons. He stated that experience proved the greatest success occurred when the animal had been dead for less than a year, had not been poisoned, and had not been infested with moths, and when the flesh had not been salted. He also discovered that the best results happened when the container was a corrugated cardboard box, especially when the task of consuming the flesh took longer than the beetle's normal life cycle. The corrugation tubes in the side of the box would be bored into by the beetle, allowing them to undergo "their change" and then return to continue the cleaning with little interruption. In his summary, Bunk noted, "Our success was not from the beginning, but gradual and after cleaning several thousand skeletons of birds, mammals, reptiles and amphibians we feel we should pass our findings along." The perfection of Bunk's system directly contributed to why KU has such a large and significant collection of vertebrate skeletons. But of even more widespread importance, this method is still used today to prepare and preserve skeletons for collections around the world.

Early on, every time one of Bunk's students left KU, they passed on by word of mouth his successful process of cleaning bones with dermestid beetles. For example, in 1922 Remington Kellogg reported success in cleaning skeletons with Bunk's process while working as an assistant biologist for the US Biological Survey in Washington, DC. Later, E. Raymond Hall carried the knowledge of the process to the Museum of Vertebrate Zoology at UC Berkeley in 1924. Hoping to see the process in action firsthand was one of the reasons that Grinnell visited KU in 1925. Initially the museum at Berkeley was slow to embrace the new process due to a preparator by the name of Edna Fisher — as the story goes, she had little interest in cleaning bones with insects and preferred the boiling process. However, in 1929, when Ward C. Russell became the preparator for the MVZ, he began utilizing the process. In 1933 Hall and Russell published an article in the *Journal of Mammalogy* about the dermestid beetle process that credited Bunk with teaching it to Hall. Subsequently, MVZ created a substantial collection of skeletons (eighty thousand cleaned by Russell) with the little insects.

In 1923 Bunk received a telegram from Dr. Alexander Wetmore, who by that time worked at the US Department of Agriculture. As the head of an upcoming scientific exploration, he invited Bunk to join them as a field collector. He told Bunk that the party would travel to the small island of Laysan, located in a chain of islands extending to the northwest from the Hawaiian Islands. Domestic rabbits had taken over the island's fragile ecosystem and they needed to be eradicated. This situation arose shortly after the turn of the century, when the guano industry began mining feces from the local bird population for commercial purposes. The rabbits were introduced to the island ecology as a food source for their workers. When the mining operation was discontinued, the rabbit population flourished, eventually decimating the vegetation on the small island. Ironically, the bird population that had created the large guano deposits could not survive without the vegetation, so several species of birds became extinct.

Wetmore impressed Bunk by telling him that the exploration party would include a nationally renowned biologist, a botanist, an ethnologist, a photographer, a herpetologist, and a rodent control expert. Not only would they eliminate the unwanted rabbits, they would survey the islands for animals, plants, and artifacts from ancient cultures. Unfortunately, a dispute arose over which institution would receive the specimens collected, resulting in the withdrawal of monies specified for funding Bunk's position on the expedition. Although Bunk never publicly complained about missing a once-in-a-lifetime opportunity to travel to a remote, exotic island in the Pacific with a party of nationally known scientists to collect birds that likely existed no other place on Earth, his greatest disappointment was not being able to join his old friend once more on a collecting trip.

Despite his feelings of disappointment for not going on the expedition, his spirits were given a boost later that fall, when T. S. Palmer, the secretary of the American Ornithologists' Union, advised Bunk that at the Forty-First Stated Meeting of the AOU, held on October 10, 1923, in Philadelphia, he had been elected as a full member of the society. That designation was considerably more prestigious than being elected an associate.

The bonds between Bunk and his former students endured throughout the years. Correspondence between them occurred regularly. They wrote him about everything in their lives: new positions or appointments, new discoveries, and births of their children. Likewise, Bunk shared events that occurred at KU. In 1924 Wetmore wrote a letter to Bunk telling his old mentor about being hired by the Smithsonian as superintendent of the National Zoological Park. Bunk was surprised to also learn that after just two months on the job, Wetmore was promoted to become an assistant secretary of the Smithsonian in charge of the National Museum, one of the preeminent museums of natural history in the country. According to Bunk, the only higher position would be the secretary of the entire Smithsonian Institution.

Not only did the careers of Bunk's boys soar, they often intersected. In

1928 Kellogg left the US Biological Survey to join his old friend Wetmore at the Smithsonian. Kellogg took the position of assistant curator of mammals, where he was in the height of his research about whales and seals, both extinct and living, and how they migrated from the warm breeding grounds to the colder feeding grounds near both the North and South Poles. During Kellogg's time at the Smithsonian he created possibly the most comprehensive collection of fossil marine mammals in the world. Whenever Bunk learned the news of such professional triumphs from his boys, it filled him with pride.

As the careers of former students were on the rise, more talented young students continued to find their way to the KU Natural History Museum. Claude William Hibbard, who originally hailed from Toronto, Kansas, enrolled at Kansas University in 1926 intending to studying pharmacy. However, in 1928 he discovered a fascination with paleontology when he signed on as a cook for a fossil field trip led by Martin to the Edson Quarry in northwest Kansas. One day after he had completed his kitchen chores, he went to the dig site to, as he put it, "find a camel and a rhino." Much to his dismay Martin offered him only tweezers to sort through rocks and soil for small fragments of fossil bones. Although Hibbard did what he was told, the slow, tedious labor got him to thinking that there had to be a better way to accomplish his task. To make his work easier, he fashioned a sieve out of window screening that he borrowed from the landowner of the property where they were digging. Water from a nearby buffalo wallow, mixed with the soil being examined, would pass through the screen, leaving behind small bone fragments. He perfected this technique over the years, which allowed him to collect an enormous amount of small fossil bones and teeth. Like others in the museum, as a sign of his acceptance, Hibbard soon received a nickname — "Hibbie."

CHAPTER SIXTEEN

An Unexpected Eviction and Well-Deserved Recognition

By and large, Bunk considered the 1920s quite personally successful. By 1930 the size of the specimen collections of birds and mammals at the KU Museum of Natural History had grown to twenty thousand and eleven thousand, respectively. The museum had achieved those numbers despite the state legislature's continued miserly control of the museum purse strings. When Dyche died fifteen years earlier, putting Bunk in charge enabled the university to manage a major museum collection with the minimal expense of the meager salary of a museum assistant. To put it in perspective, in 1929 Bunk earned only $2,400 per year (about $34,600 today when adjusted for inflation), about half of what a member of the faculty would make. In addition to the growth of the collection, Bunk was especially proud that his former students continued to enjoy considerable professional success, and each year he continued to send forward new waves of young and talented scientists.

As the decade opened, Bunk continued to contribute to the body of knowledge of ornithology. In the spring of 1930 he reported to the American Ornithological Society the first example of a starling recorded in Kansas. The starling was first introduced into the United States in 1890 in Central Park in New York City by Eugene Schieffelin, who wanted all of the birds mentioned in Shakespeare's works to reside in the New World. Despite his intentions, it took four decades for the species to reach halfway across the continent. In February a frozen specimen of a starling was discovered in a silo just west of Bronson, Kansas, in Allen County. A KU student by the name of O. Ireland delivered the bird to Bunk, who added it to the collection.

The 1930s provided many changes in Bunk's world. When he turned sixty years old he realized he was beginning to physically slow down. Coincidentally, the state economy began to slow down also, as the Great Depression crept across the country. Repeated years of extreme drought combined with over farming created a Dust Bowl, in which fragile soils were blown to the East Coast in the form of mile-high walls of dust. Moreover, Bunk lost another longtime friend, and his beloved museum would experience an unexpected hibernation.

Midmorning on January 15, 1931, Bunk received word that his old friend Handel Martin had finally succumbed to the cancer he had battled for two years. In spite of his initial diagnosis, the energetic Martin had continued to excavate fossils in the field, even after he became frail and emaciated. In the spring of 1930 Martin underwent cancer surgery, from which he never fully recovered. Although he never worked in the field after his surgery, Martin continued to visit his fossil workshop in Dyche Hall on a daily basis. The pain and weakness that resulted from the surgery made it impossible for him to walk up the stairs to the third floor of the museum. To remedy that, several of his students rigged a chair to carry him up in the morning and down in the evening, allowing him to spend the day cleaning and fitting bones of his prized fossils.

During Martin's illness, Bunk made a point to visit his old friend on a regular basis. They reminisced about their days in the field—hunting at Bunk's cabin or their trips to western Kansas. The two old warhorses of the museum enjoyed their time together, sharing their memories and gleefully recounting how they successfully protected their collections from the ambitious faculty. In addition, Bunk kept Martin apprised of his current field trips, like the summer field trips to New Mexico and Arizona, in the summers of 1929 and 1930.

After Martin passed away, Bunk planned one more field trip to the southwest. In June of 1931 Bunk and five students departed from Lawrence to collect near Phoenix and Tucson, Arizona, and along the Mexican border. They successfully returned with three hundred bird skins, two hundred reptiles, and four hundred skeletons of various mammalian forms. They also brought back a transparent lizard, a live Gila monster, a rare specimen of a rattlesnake, and a large land turtle with horny scales. That 1931 collection trip to Arizona proved to be Bunk's last extended expedition. After that year, Bunk's name only appeared as a collector of record for specimens taken from around the Lawrence area. The rigors of field trips in faraway places had taken their toll. His aging body no longer allowed him to participate in strenuous camping trips like the ones he had enjoyed with Theo and the unforgettable trip to Alaska. According to Fedalma, when Bunk could no longer take on the more strenuous field trips, "he was a changed man; he seemed to feel that his efficiency and usefulness had been definitely impaired."

With Martin's demise, the question arose as to who would replace him as the curator of the vertebrate paleontology collection. The austere financial conditions of the state created a dilemma for Henry Lane: he wanted to hire an academic, but only had the budget for a preparator. Where could he find capable men like Bunk and Martin who were willing to accept the position for so little? Lane reached out to Curtis Hesse, a former graduate student who had worked for Martin. Lane hoped that Hesse, then a doctoral student at UC Berkeley, might work for less while earning his doctorate at KU and could earn better pay when the state funding returned. Hesse turned him down because of concern about the long-term funding problems at the museum. Lane shared with a number of colleagues across the country his fears that he would be limited to hiring only a preparator, rather than a paleontologist. Having no success in finding a qualified paleontologist and believing that it would be years before the salary budget would increase, Lane accepted the inevitable and assumed the role of "acting curator in charge of the fossil collection."

When word got out that the natural history museum at KU was suffering financially and had lost its longtime, well-respected curator of fossils, it was not surprising when individuals and institutions began looking for opportunities to take advantage of KU's distress. One rumor circulating involved the son of celebrated fossil hunter Charles H. Sternberg. In 1928 George F. Sternberg joined the faculty of Fort Hays State University and was trying to create a fossil collection and museum in Hays, Kansas. That caused some to speculate that a possible solution to KU's dilemma would be to incorporate or combine somehow the two state-supported museums. Even Hesse, who was far away in California, had heard the rumors and expressed his support to Lane in his January 27 letter, writing, "I rather feel that the fears of the man from Hays are groundless — Dr. Lindley and you yourself would never see such a thing go thru." In addition, in October of 1931 Barnum Brown, by then the curator of fossil reptiles at the American Museum of Natural History in New York City, wrote Lane about trading some of his fossils for Bunk's mammoth sea serpent from 1911. It was common practice for museums to buy, sell, and trade specimens to fill out certain curatorial needs within their collections. However, directors of museums had to be careful not to undermine the value of their collections. Trading such a significant specimen as Bunk's mosasaur might have righted the financial ship of the museum, but at the expense of seriously diminishing the collection. Such a solution was like a farmer selling his seed corn. Lane smartly turned down the offer from Brown, saying that short-term budget cuts had only temporarily slowed the preparation of the mosasaur.

In late fall of 1931, when Bunk opened his recent issue of the *Condor*, he grinned ear to ear. One of his former students, Lawrence V. Compton, had been published in the prestigious periodical. Compton had just graduated from KU in 1929 and was working on his master's degree at the University of California, Berkeley. His article announced a new record: the first-time evidence of white-winged scoters in Kansas. Normally coastal birds, they are seldom sighted inland and then only on rivers and lakes during times of migration. Compton reported how the first was killed on November 21, 1927, by William Sanderson. The following year Pug Saunders shot four more sometime in November of 1928. The approximate location of all of these birds was a few miles up the Kaw River from Lawrence. The article concluded with acknowledging "Mr. C. D. Bunker . . . for permission to report this record." In truth, for Bunk it was always more about encouragement than permission. Here was another example of how Bunk easily could have published that record, but he wanted his "boys" to publish and get the credit, as the prestige and experience associated with publishing was more important to their careers than to his. The wisdom of that approach by Bunk was confirmed as Compton went on to become the chief biologist of the Agriculture Department's Soil Conservation Service.

For nearly two years after the death of Martin, the bleak financial situation at the university continued to hang over the KU Natural History Museum. Just when matters looked as if they could not get any worse, an event occurred in Pittsburg, Kansas, that seriously jeopardized the future of the KU museum and its various collections. When part of a ceiling collapsed in a building on the campus of the Kansas State Teachers College, the board of regents directed the state architect to do a survey of all state university buildings. During the examination of Dyche Hall, the inspectors found that the concrete floors had begun to buckle. A cost-cutting decision during the original construction of Dyche Hall, to utilize wooden timbers and barbed wire to reinforce the concrete floors, rather than steel beams and rebar, was now causing the structure of the building to fail, and the state architect condemned the building. On November 30, 1932, the regents ordered Dyche Hall vacated and closed. As a result, the faculty and staff, plus most of the collection, were relocated to various buildings around campus. Unfortunately, because the mounted animals in the panorama were so firmly attached to the landscape, they could not be removed from the museum. Bunk and the staff attempted to protect the valuable mounts from dust by draping them with sheets and tarps. Bunk's office was moved to the animal house on the south slope of Mount Oread. With the Great Depression drying up both public and private funding sources, repairs to the old building were doomed for the foreseeable future. As a result, the great Dyche Hall, with its *Panorama of North American Mammals*, remained shuttered through the remainder of the decade.

Although the museum was closed to the public, the staff still managed to keep busy. A news release from KU in July of 1935 reported that mold had attacked a portion of the study collection stored in the basement of Hoch Auditorium. Bunk estimated that the restoration task would involve thousands of skeletons, mammal and bird skins, and even some mounted birds that had been stored and sealed in boxes. The mold developed as a result of a damp spring combined with a hot summer. To avoid a reoccurrence, the cleaned specimens were moved to the unfinished top floor of Dyche Hall, where they could be kept dry.

By the 1930s Bunk's long career began to receive recognition. The September 1934 edition of the *Condor* included a glowing article on Bunk's career and contributions to vertebrate zoology. He was acknowledged not only for work assembling a substantial collection of birds and mammals but also for his work identifying promising students and training them for careers in zoology. Referring to his ability to wisely select "human material," the article described him as clairvoyant, noting that "without a conscious knowledge of pedagogies his results with students have been those of a really gifted teacher," and "with no thought whatever of reflected glory he fairly gloats over their every contribution to the literature." Bunk tried to downplay the recognition, but greatly appreciated the sentiment. It should come as no surprise that the editor of the *Condor* at that time was Joseph Grinnell, and Bunk's former student Jean Linsdale was the associate editor.

In addition to Bunk's peers, the university also recognized his career. In the fall of 1935 Bunk was asked if he would sit for an interview for the *Graduate Magazine*, a periodical published by the University of Kansas Alumni

Photo of Bunk taken circa 1934 (University of Kansas Natural History Museum records, University Archives, RG 41/0 Photographs, Kenneth Spencer Research Library, University of Kansas Libraries)

Association that routinely contained articles about the university and distinguished faculty and alumni. Wanting to acknowledge Bunk for his time at the university, the magazine assigned a senior named Catherine Penner to write the article. For the interview, she visited Bunk's office in the animal house on the morning of December 12, 1935, his sixty-fifth birthday. Given that his musty old office in the animal house was off the beaten path, which suited Bunk just fine, he met her on campus and escorted her down the steps on the south side of the hill past the university power plant.

As Penner recounted, Bunk directed her to what he referred to as "the best seat we have to offer." She described it as a "rather charred old chair with a ticking-covered pillow for a cushion." With a twinkle in his eye, Bunk explained, "We reserve it for the ladies." Once they were seated opposite each other, Penner noticed that Bunk seemed uncomfortable answering questions about his own accomplishments. However, as soon as she steered the conversation to his former students, he relaxed. In fact, Bunk "fairly bubbled over with excitement, while he fingered a large pile of letters he had received from them congratulatory of his sixty-five years of achievement." Perceptively, Penner began to better understand Bunk's role at the university, whereby she described him in the article as the "Kingfish of the Museum."

Although Penner only met briefly with Bunk, with the help of his former students she ably painted a poignant picture of a man with an extraordinary history at the university. Penner borrowed those letters from Bunk's students, thinking that they might offer some help in organizing her article. But after reviewing them, she wisely decided that the best way to portray the esteem in which Bunk was held was to use the actual heartfelt words of his students, verbatim. Those letters are as follows, just as they appeared in the *Graduate Magazine* in January of 1936.

Some of the Messages to an Old Friend

Patient Strength to Stay

I read a poem once about a man
Who built a bridge across a chasm wide;
I used to fear that I might have to come
That way alone, some darkening even-tide,
And through the flying spray and roar might fall
Before my gropings found the other side.

If I could write a poem I should choose
Not one who builds a bridge and goes away;
But one with furrowed cheek, and beetling brows
Above his eyes of granite gray,
Who tends the bridge youth's feet must come along,
The man who has the patient strength to stay.

Maye Hooper Leonard, Student, Dept. Zoology, Lawrence Kans.

The care with which you have kept your catalog records has been unusually exact; your originality and labor in the erection of the panorama will stand as a remarkable and unique contribution to museum technic [*sic*]; your method and long success in the preparation of skeletons until you now have perhaps the largest collection of such specimens in this country, if not in the world, is a monument of which you have every right to be proud; and finally, the influence you have had upon the careers of so many men now in active scientific service marks you as one of the most important figures of your generation in your field.

H. H. Lane, Prof. Zoology, University of Kansas, Lawrence, Kans. (Director of the University of Kansas Natural History Museum)

When I think back, in terms of years, we have known each other a long, long time, though I always remember our work together as though it took place yesterday. The work that we did together represents one of the happiest periods of my life, and I treasure it as experience that has been invaluable to me in after years. Many things have come since to both of us, a part of them good and a part perhaps not wholly pleasant, but our friendship has continued unchanged in spite of our separation in space.

Days afield such as you and I have had bring men into a closer contact and into a closer mental affinity than is possible in any other way.

Alexander Wetmore, '12, Asst. Secy, Smithsonian Inst.; Head, U.S. Nat. Museum, Washington, D.C.

Perhaps I am in a very good position to see your fine work in a proper perspective, for, as you remember, I was among the first to come under your influence in the old days of Snow Hall.

I see before me a long list of men and women who, starting with the inspiration that you had to offer, went on to assume prominent positions in administrative, teaching and research in the scientific world.

The museum at the University of Kansas will always stand as one of your outstanding accomplishments, for it is largely through your foresight and labor that it has become the depository for research materials that it now is.

The materials library that you have built up will be useful long after the showy exhibits have disappeared.

Leverett Allen Adams, '03, g '06, Prof. Zoology, University of Illinois, Urbana, Ill.

The little incidents and happenings about the museum and in the field are the high-lights of those pleasant days. Trapping for "coons" on the Wakarusa — looking for hermit thrush — trying to catch Synaptomys — working half of the night on a collection of fresh specimens — catching rattlesnakes and copperheads — shooting whip-poor-wills — hunting great-horned owl nests. Such incidents, though they may have seemed insignificant at the time, are seldom, if ever, forgotten. You enter prominently into each of these memories.

Personally, I feel that most of my inspiration and interest in my chosen field of study was initiated by you there in the old museum and in the field about Lawrence. Without your help and encouragement, I could not have carried on.

R. A. Stirton, '25, Curator of Paleontology Collections, Museum of Vertebrate Zoology, University of California, Berkeley, Calif.

In this day of huge and startling accomplishments and the sordid struggle of commercialism, we are prone to pay homage only to men who are exercising great power, regardless of their history. It is good for us, sometimes, to pause for a moment and give consideration to the fact that the honor of the nation, and of every part of it, and the happiness of our people, and every individual member of it is dependent upon straightforwardness of character, efficient industry and unselfishness — all this I have in lively remembrance of you, as I think of the long and faithful service rendered by you at the University of Kansas.

Ben Bolt, Kansas City, Mo.

There must be great satisfaction in the knowledge of a life work well and conscientiously done. Like dozens of other men who have worked with you in your "smelly old workshop" I shall always be eternally grateful for the boost you gave me over the rough places. Not everyone would have been interested in the problems of every student in the shop, regardless of own interests.

W. S. Long, g '19, Ph.D. '30, Field Naturalist, Zion National Park, Springdale, Utah

I recall to mind our conflicts with Aunt Carrie Watson [longtime KU librarian] about the withdrawal of books necessary for the identification of specimens. Then we were favored from time to time with the visits from the one and only L. L. Dyche. His seemingly infallible memory and inexhaustible stock of anecdotes about how he came to collect each specimen in the museum afforded entertainment to unseen listeners. I can still visualize big Doc Sundwall and Dr. Coghill sticking their necks out from behind convenient cases in the basement while he was holding forth on his experiences. Then there was that early morning visit from the justly famous Board of Regents about the time of their "no smoking" and "no swearing" edict. On this occasion I believe I was expressing my sentiments about some mice that had been swept off the table by the janitor, when I turned and beheld them in all their glory. They must have felt that this situation warranted an exception.

Seriously though Bunk, I owe much to you, for if you had not given me the chance to work in the museum I am sure that subsequent events would have worked out much differently than they have.

Remington Kellogg, '15, g '16, Asst. Curator of Mammals, U.S. National Museum, Washington, D.C.

I am not merely grateful for what you did for me while I was in school, but there is the addition of a philosophy — a strong contempt for the superficial, a great respect for the practical, and about all, a belief in square dealing — which keeps alive in my consciousness, and I am grateful to you for planting it there.

Harry C. Parker, '30, Director, Museum, Worcester Nat. History Society, Worcester, Mass.

Not forgetting your personal accomplishments as a naturalist afield and also the success of your curatorial attentions to the University collections under your charge, I have in mind at the moment your signal success as a teacher. In trying to analyze the means which you employed, I think there was included: (1) a period of apprenticeship to determine an aptitude; (2) encouragement for all tasks well done; (3) suggestion, as opposed to direction, of problems possible for the student to solve; and (4) a disclaimer of any part in the ultimate accomplishment to the end of advancing the student's own interests. This system of teaching, only lately recognized by educational authorities as one of the most successful methods of instruction, reflects the sound judgment you have brought to bear on problems of the Museum in general. Certainly the wide distribution of your "boys," now actively contributing to the sum of knowledge in the field of natural history, place[s] you without a peer.

E. Raymond Hall, '24, Curator, Mammals, Museum of Vertebrate Zoology, Berkeley, Calif.

We know that you have done much for K.U. students in the past and your work at the University will be more and more appreciated as the years roll by. That skeleton collection, the best in the world, that you have gathered together will some day [*sic*] be fully recognized and your forethought in early starting the work will draw much praise. In my opinion that one collection is great enough to repay the University for your forty years service and could well stand as a fitting monument for all your efforts.

Charles C. Sperry, '18, U.S. Biological Survey, Denver, Colo.

The younger generation owes you a debt that they never can repay, and we older ones also are under a tremendous obligation to you. May you live long and continue to prosper in health, in personality and in accomplishment. To me, you are an outstanding example of the best that there is in practical scientific education. You not only instruct the student, but you instill a love for science that will continue to be a powerful and stimulating force long after you are gone.

Dr. R. L. Sutton, Big Game Hunter and Author, Kansas City, Mo.
(Reprinted from Graduate Magazine, *University of Kansas, volume 34, number 4, January 1936)*

When those letters began arriving prior to his sixty-fifth birthday, the genuine feelings expressed by his former students truly moved Bunk. But this would not be the last time his "boys" showed their loyalty and affection for their old mentor. A decade later, as his career neared its end, his students would once again rally around Bunk.

About the same time that the article appeared in the *Graduate Magazine*, Bunk and William R. Green, a friend who owned a local hardware store, began gathering live animals to create a private zoo. They located it on a sizable parcel of land near the banks of the Kaw River that included a small lake and several cabins. Green sought out Bunk to collect animals and build cages for the new zoo. The first animals they collected were two coyotes they trapped on the property. Soon afterward they opened the zoo to the public. According to Will Green's great-nephew, "In time, the Zoo had four bears, a 7′ alligator and two smaller alligators, four or five coyotes, a pet crow, six raccoons, two red foxes, a monkey, porcupines, a mountain lion, a Gila monster and other animals, mostly local animals, that were kept in cages." Green and Bunk hoped the city would eventually take over maintenance of the zoo, but that never occurred. Without public support, the zoo struggled for decades. By 1951 the zoo had only one monkey, which died when a great flood devastated the site, finally closing the zoo.

Although naturalists and field zoologists care deeply about the animals in nature, they also recognize the importance of growing their collections to benefit science and the long-term life of the planet. Bunk's involvement in the zoo was a bit of a departure from his life's work, in that it is the only record of his working with live animals. Up until then his focus was hunting and killing specimens for the collection. Whether there is any significance in that change of activity will never be known for sure; however, after the death of his eldest daughter, Fedalma, in the 1990s, several old photographs surfaced from her personal effects raising the question as to how Bunk actually felt about killing all those animals. On the back of a photograph of Bunk was a note in Fedalma's handwriting saying, "I regret every specimen I ever took. They wanted to live too." She signed it "C. D. Bunker." That note is the only indication of any remorse felt by Bunk for all of those years of collecting. However, because Fedalma was so emotional about animals it is possible that the quotation was an expression of her personal feelings rather than of her father's.

* * *

As it came to pass, Martin's replacement as curator of the paleontology collection rose from within the ranks of KU. While Lane patiently waited for sufficient funding to return to the museum budget, he utilized the services of a talented student assistant to supplement his own service as acting curator in charge of the fossil collection. Claude Hibbard helped Lane while he earned his bachelor's degree in 1933 and his master's in 1934 in zoology. Following graduation he worked briefly in Kentucky as a wildlife technician for the National Park Service. After Lane came up with the funding to hire Hibbard, he brought him back to KU as the assistant curator of vertebrate paleontology in 1935. Hibbard served in that position until 1941, when he was appointed curator of the collection and made assistant professor of zoology.

Bunk was pleased to have Hibbie back at KU in 1935; he especially liked the idea of hiring "one of their own." In addition, paleontology would benefit from his scholarship. Hibbard thought that paleontology was inseparable from zoology and geology; all were indispensable parts of a whole. As a testament to that belief, beginning in 1936 Hibbard commenced a thirty-nine-year career of studying fossils of Meade County, the county west of where Bunk and Theo had camped along the Cimarron River in 1911. Throughout his career Hibbard's research focused on how dramatically the Kansas climate had changed over millions of years. His noticed how long periods of cold were followed by extended periods of heat. Proof of those extreme climate changes came from finding fossils in various strata of soil and rocks belonging to animals that could not typically cohabitate together. For instance, one rock strata contained animal life that thrived in cold weather, indicating a glacial period, whereas in another layer he found fossils that were the early relatives of alligators and elephants, which signified a warmer climatic period.

In the fall of 1936 James M. Sprague joined the cadre of young scholars who came under Bunk's influence. Initially from a once-prominent family in Kansas City that fell on hard times during the Great Depression, he needed to work once he enrolled at the University of Kansas. He was thankful that he could work at the natural history museum and teach an anatomy lab. Though he was paid next to nothing and worked long hours, Sprague recounted how he thoroughly enjoyed his time at the museum. His experience at KU encouraged him to believe that he would have a career as a scientist and follow in Bunk's footsteps in a career as a museum curator.

However, World War II changed that. When he was ready to leave graduate school at Harvard for a position as an assistant curator at the Field Museum in Chicago, the War Department determined that such a position was not high priority enough to keep him out of military service. Looking for another option, he found openings at Johns Hopkins University for medical school, as doctors were in demand for the war effort. As fate would have it, this change in his professional career worked out to his advantage, as he became a successful neurosurgeon. In spite of this career change, his time at KU had been influential to his career and Sprague continued to appreciate his time in the natural history museum.

Before going off to graduate school at Harvard University, Sprague received an invitation from Bunk to sign what was referred to as the "buffalo bone." This much-revered old relic, an actual scapula from a bull bison, measured almost two feet in length. Signatures and dates were written all over it, like a cast for a broken arm for the most popular kid in school. Even though Bunk was never in a position to offer grades to the students to acknowledge their achievements, he was able to let them know they had "arrived" by inviting the gifted ones to sign the bone, a testament to the world of the student's demonstrated talent as a naturalist.

Many years later a photograph of the bone showed the prominent notation, "Picked up at the cabin Nov 1927," indicating that the ritual of signing the bone first became a tradition in the late 1920s. Although that date was after the time of Alexander Wetmore, Remington Kellogg, E. Raymond Hall, and William Burt, their names are inscribed as a result of Bunk offering the honor of signing to early members of the "gang" whenever they returned to their alma mater. In Sprague's own words, "Eventually I was asked to sign along with these men (and two women) and did so proudly with the knowledge that I was now a qualified naturalist."

By 1938 Kellogg had joined the US National Museum at the Smithsonian as the assistant curator of mammals. On a cool, crisp November morning Kellogg returned to his office to find a letter on his desk. It was from a former KU student, an attorney practicing law in Wellington, Kansas, a small town in the southeastern part of the state. H. W. "Goody" Goodwin explained how he had graduated from KU in 1921 and had collected birds and mammals around Lawrence for Bunk during his time in law school. He had always thought that Bunk deserved the title of curator, rather than assistant curator. He confessed that he had wanted to do something about what he called an "injustice," but he had always put it off until he was reminded again by the 1936 *Graduate Magazine* article honoring Bunk for his long career at KU and the museum. Despite his procrastination, Goodwin now had a plan to get Bunk the recognition he deserved. Using the names of the former students in the magazine article as his mailing list, Goodwin wrote letters to a handful of them to enlist their help by lobbying the board of regents to change Bunk's title. Goodwin admitted that Bunk knew nothing of this plan and that he wasn't even sure if Bunk would approve of the action, but he asked them to write letters of support nonetheless. He wanted their letters addressed to the board of regents but asked that they send them to him. He in turn would use the letters to approach the regents, many of whom he said he knew personally.

Kellogg immediately wrote a letter to Goodwin supporting the idea of giving Bunk his recognition but cautioned Goodwin that going to the board of regents might cause problems for Bunk in the long run. Kellogg explained that Chancellor Lindley was listed in the university catalog as the director of the museum and Lane had the title of curator of the museum. Kellogg suggested that Goodwin discuss his proposal with both of those men directly, rather than the regents. Coincidentally, Goodwin got the same advice from Hall, Stirton, and Wetmore.

Goodwin's initial idea of approaching the regents emanated from his background in politics and existing relationships he had with the board members. However, he immediately recognized the wisdom of not circumventing the lines of authority. By the time he notified Bunk's former students of the change in strategy, the situation at the University of Kansas had unfortunately changed. During the time that this correspondence was going back and forth, Lindley announced that he was going to retire, effective June 30, 1939. Considering that one of the two key players in such a decision was going to be replaced, Goodwin's well-intentioned movement to recognize Bunk with a new title promptly stalled.

Six months later Lane informed Bunk of the name of the person replacing Chancellor Ernest Lindley. Much to Lane's surprise, Bunk let out an uncharacteristic yelp and slapped his knee. Bunk explained that his unusual reaction resulted from the news that the incoming chancellor was a former student and one of Bunk's boys. His time at KU preceded Lane's arrival. Bunk was always proud when his former students achieved success, but he never dreamed that one might actually become chancellor of KU. As the

Bunk circa 1940 (University of Kansas Natural History Museum records, University Archives, RG 41/0 Photographs, Kenneth Spencer Research Library, University of Kansas Libraries)

eighth chancellor of the University of Kansas, Deane W. Malott would be the first chancellor who was both a native Kansan and a KU graduate. Born and raised in Abilene, Kansas, Malott was the son of a successful banker who was close friends with the Eisenhower family, and he had majored in economics at KU and earned an MBA from Harvard. Bunk told Lane about how Malott spent his entire freshman year comparing every bird specimen in the collection against the museum catalog. Bunk joked that this would be the first chancellor who actually had worked for him.

The 1939 Kansas legislature appropriated the final amount to finish restoring the KU Natural History Museum, and in the fall of 1939 Lane hired Klaus Abegg, a Swiss taxidermist, to clean and restore the mounts in the panorama. Although his résumé of museum clients included the American Museum of Natural History in New York and the Field Museum in Chicago, the KU museum impressed him. Abegg was quoted as being "delighted; he had expected the usual college museum, crowded into the top floor of some building." As the decade came to a close, the museum community looked forward to the reopening of Dyche Museum. Lane and others were hopeful that they could open at least one floor by graduation of 1940.

CHAPTER SEVENTEEN

The Museum Reawakens, Bunk's Boys Come to the Rescue

Although hopes had been high for reopening Dyche Museum in time for commencement in 1940, it would not come to pass until a year later. In early 1941 Bunk and Hibbard proudly began the colossal task of moving everything back to the museum building. Meanwhile, Lane prepared for a reopening and rededication ceremony. The grand weekend of events for the university included the rededication ceremony for the reopening of Dyche Hall on Friday, the celebration for the seventy-fifth anniversary of the founding of the University of Kansas on Saturday, and commencement exercises for the class of 1941 would cap off the weekend on Monday. With rain included in the weather forecast for that Friday, they moved the rededication ceremony for Dyche Hall to the auditorium of Old Fraser Hall. Wetmore, as the featured speaker, spoke eloquently to the more than four hundred guests in attendance. During his remarks he spoke about the value of the museum to society. "But let me impress upon you the very pertinent fact that these displays, attractive tho [*sic*] they may be, are not the most important properties of this museum." He went on to explain that even though the exhibits are what the public sees and appreciates, it is the work of men like Bunk, working behind the scenes that make it all possible. "They form a portion of the valuable properties of our university, and constitute in considerable part the dynamic force of this museum." The guests of honor for the affair were the widow and daughter of Lewis Lindsay Dyche. After the ceremony, hundreds of guests privately toured the renovated Dyche Museum and it finally reopened to the general public on the day after graduation.

The long-awaited reopening of Dyche Museum had been delayed for nine years for two reasons. First, a lack of adequate funding surrounding the economic depression of the 1930s made public funding for projects of this kind difficult to accomplish. Monies for the repair arrived piecemeal during the nine-year period, from the state of Kansas as well as through federal funding from the Public Works Administration and Works Progress Administration. Just to make the building habitable, all existing floors were replaced with newly poured concrete floors and added reinforced steel beams. In addition, they excavated the basement (ground floor) and provided new concrete footings to support the weight of new rebar-reinforced concrete. The second reason for the delay of the reopening was that they increased the scale of the job. Since the museum was already displaced due to the repair work, the museum administration took the opportunity to expand the exhibit space by adding an entirely new floor. By extending the small mezzanine over the front lobby toward the panorama, they created a floor between the first floor (panorama) and the second floor (exhibits). Although the changes created considerably more exhibit space for the museum, it lowered the ceiling to fourteen feet in the viewing gallery while leaving twenty-four-foot ceilings in the panorama and the entry.

Moving the paleontology exhibits from the third (top) floor of Dyche Hall to the basement also occurred during the remodeling of the museum. As it turned out, having the mass and weight of the fossil collection on an upper floor had significantly contributed to the bowing of the poorly reinforced floors. The nine-year renovation cost $125,000, two-thirds more than the original price to build Dyche Hall. Despite this appearance of being "penny wise and pound foolish," the functionality of the museum's classroom and exhibition space greatly increased with the 1930s renovation.

Shortly after Malott took office at KU, he discovered that Bunk still held the restricted title of "assistant curator in charge." Recognizing Bunk's abilities and years of service, Malott felt that Bunk deserved the full title of "curator of modern vertebrates." The chancellor made the announcement to coincide with the grand opening and rededication of Dyche Hall. Bunk was ecstatic when he was told of the new title. After all those years of institutional inertia limiting him to the lesser title, it was heartwarming to finally be recognized by the university's top administrator.

In March of 1942 the announcement came down that Bunk would be going to work half-time for half the pay. Bunk never complained, just like when his pay was cut during the 1930s (his pay in 1930 was $2,400 for a twelve-month appointment; by 1942 he was earning $2,200). But his boys were unwilling to sit idly by. Once the word got out, Chancellor Malott began receiving letters from former students saying that what KU was doing to Bunk was disgraceful. The first letter to arrive came from H. W. Goodwin, dated March

10, 1942. He attached the file of letters from his 1938 campaign to get Bunk the title of curator. In addition, he had heard that Bunk was being reduced to half-time and half salary on July 1, 1942, and understood that Bunk was in "rather desperate straits and is making arrangement to dispose of his entire personal library." Immediately, Malott replied by letter saying that Bunk had already been promoted to curator shortly after Malott came into office. He also explained that all state universities in Kansas have the same retirement policy: upon reaching seventy years of age, the employee goes to half-time with half pay. As a result, Malott had no control over changing the plan for Bunk. He offered that even though KU's plan was generous when compared to other universities, the problem for Bunk was more about his low salary than a stingy retirement plan. A few days later the chancellor received a letter from his friend Vic Housholder about Bunk's plight. In his reply to Housholder, Malott acknowledged that there was apparently quite a campaign on Bunk's behalf, because he was "getting several letters a day from [Bunk's] old friends."

Eventually, Bunk's boys would resolve his retirement situation. On June 27, 1942, William Burt, from the University of Michigan Zoology Museum, wrote to Malott on behalf of his former mentor and offered a constructive suggestion for how to supplement Bunk's retirement income. Burt had contacted dozens of Bunk's former students and asked them to personally make yearly contributions to an endowment that would provide "income to supplement Bunk's salary for the rest of his life then be turned into a scholarship or research grant." The kicker to what Burt called the "Aid Bunk" program was the request that the chancellor "handle it in a way that Bunker does not know where it is coming from." With the chancellor's agreement, twenty of Bunk's former students began sending their checks. Contributing members of Bunk's boys included Leverett Adams, Lawrence Compton, Louis Coghill, Ted Down (serving in the US Army), E. Raymond Hall, Curt Hesse, Claude Hibbard, John Eric Hill, Remington Kellogg, Jean Linsdale, W. S. Long, G. C. Rinker, Jim Sprague, Ruben Stirton, Dix Teachenor, Otto Tiemeier, Alexander Wetmore, Jim Whitaker, Ted White, and William Burt. They raised $400 that first year (equivalent to over $6,300 today) and agreed to keep some in reserve for future years.

On August 10, 1942, Malott sent Bunk a letter and a check for twenty-five dollars. In the letter, he said, "The enclosed check is the result of a bit of money in the Endowment Association which I am privileged to send to you. It is sent because of the unhappily small half-pay allowance under which you are now serving." Even though the money raised for Bunk's retirement was modest by today's standards, it was still an incredibly generous amount for the 1940s. Moreover, the idea of contributing to a retirement fund for a former teacher or mentor is extraordinary and would be virtually unheard of today. Such a gesture speaks volumes about their loyalty to and affection for Bunk. On June 26, 1944, Burt wrote to all of Bunk's former students that the retirement fund for Bunk needed replenishing. Burt closed the letter saying, "From what I gather, the $25 a month that Bunk gets from this fund is just enough to make the difference between hard times and a bare existence." In the end, much to the credit of Chancellor Malott, who agreed to keep confidential the source of those funds, Bunk never knew that his retirement supplement came from his "boys."

After Bunk retired, the day-to-day operation of the museum continued like before, except Bunk worked a shorter day. Initially Bunk had reservations about whether retirement would suit him. As one of those men whose identity was defined by his work, all of his self-worth was tied into the museum, nature, and students. Fortunately, his half-time position allowed him to stay connected to what defined him and after some time he began to enjoy retirement. In fact, he recognized and even appreciated that he had neither responsibilities nor authority at the museum, as those had been passed on to the next generation of museum men.

Throughout Bunk's time at the natural history museum, the faculty, staff, and students always enjoyed a collegial atmosphere, apart from those early turf wars. Everyone worked hard but always had time to celebrate personal events and professional accomplishments. One example of the camaraderie occurred annually on Bunk's birthday. Although this tradition had never previously garnered any publicity outside the museum, the *Graduate Magazine* featured a small article covering his seventy-second birthday in 1942. As was the custom, shortly before noon Bunk and the students gathered at the foot of the museum staircase. He challenged all comers to race the stairs of Dyche Museum, from the basement all the way to the top floor. Laughter and good-natured ribbing echoed up the museum's terrazzo staircase that morning. According to the article, "his speed was too well known for any acceptances." With no takers, Bunk proceeded up the stairs by himself at a leisurely pace while the students and staff rooted him on. Afterward, the students honored him with a luncheon at the museum with faculty and staff.

Two years later, in 1944, the museum experienced a significant changing of the guard. When Professor Lane turned sixty-five, a university policy required him to step down from administrative duties, though he continued to teach and do research. That meant the natural history museum needed to look for a new director. Malott conducted a thorough search and selected E. Raymond Hall. Hall had been working at UC Berkeley as the curator of mammals and acting director of the Museum of Vertebrate Zoology after Joseph Grinnell died. Both his graduate degrees from Berkeley and his administrative experience made Hall the perfect candidate to lead the KU Natural History Museum. Knowing that two of his boys were now serving in high places at the university gave Bunk great personal satisfaction and pride.

In 1945, when Bunk turned seventy-five, KU retirement continued his half pay, but there was no expectation for him to work. He was not ready to say goodbye to the museum, so he continued to spend time at Dyche Hall. As part of his established daily routine, he usually arrived at the museum between eight and nine o'clock. After checking to see if he had any mail, he would spend the morning reading current scientific journals or assisting students in the workshop.

Bunk's daily visits to the museum also allowed him to engage in his favorite activity: keeping up to date on the successes, both personal and professional, of his former students. For instance, in 1945 news reached Bunk that Wetmore had been appointed as the sixth secretary of the Smithsonian Institution. He was overjoyed that one of his first boys had attained, as Bunk described it, "the highest scientific position a man can attain in this country or possibly the world."

Sometimes news came by letter or telephone, but occasionally it came in person. Whenever his boys returned to their alma mater, they always stopped by the museum to visit with Bunk. One such meeting occurred in the spring of 1947, the day Kellogg was coming to town. When Bunk awoke that morning, like most mornings, the stiffness of old age made physical movement difficult. However, as soon as he remembered what day it was, no physical pain could keep him down. Kellogg was then in his mid-fifties and had been promoted several years before to curator of mammals at the US National Museum at the Smithsonian in Washington, DC. Although he was in the prime of his career, Kellogg was never too busy to visit with his old friend and mentor whenever he was in Lawrence. Bunk and Kellogg met that morning at Dyche Hall, greeting each other in the familiar, comfortable museum workshop. Sitting on stools at a worktable, they visited quietly for more than an hour, reminiscing and telling stories about the old days. When these close friends concluded their conversation and said their goodbyes, Bunk headed home for dinner.

During their time together, it occurred to Kellogg that the story of Bunk's life needed to be preserved. With Bunk now in his seventies, there was legitimate concern that the events of his life might be lost if not recorded then — a simple man like Bunk, who had always maintained a low profile, would have little mention in public records to chronicle his life. Unlike Professor L. L. Dyche or Chancellor Francis Snow, both men who had published extensively, held high office, and had national reputations, Bunk's correspondence files and press clippings could best be described as scant. After his meeting with Bunk, Kellogg went directly to Hall's office and announced that they should do something to preserve the history and memory of their old mentor. With Kellogg feverishly dictating what he remembered from that conversation with Bunk, Hall made detailed notes that later became the basis for a published memorial for their old friend and colleague.

When Bunk drove home from campus, he thought back fondly on his conversation with Kellogg. Bunk and Clara then lived on the western edge of Lawrence. In front of his house, the pavement ended and the street continued on to the west as a gravel county road. As a semirural neighborhood, the nearby homes were situated on larger lots of several acres. Most residences had fences to accommodate their milk cows, horses, and chickens. This was the first time that Bunk had not lived in the southern part of Lawrence, but he enjoyed it because it felt more rural, like the countryside that he so enjoyed in his younger days.

The crunching of the tires on the gravel driveway ceased when the car came to a rest behind the house. As Bunk entered the back door to the kitchen, the spring on the screen door drew it firmly to the frame with a muffled crack. He sat at the kitchen table covered with a worn oilcloth as Clara prepared him a ham sandwich and a glass of milk. Bunk's eyes twinkled as he shared with Clara his visit from Kellogg. He talked with the pride of a father when he recounted what was going on in Washington, DC, and in the life of one of his boys.

Bunk's conversation with Kellogg provided immeasurable joy for the remainder of the afternoon. Although these visits with his old protégés were important to Bunk, they never seemed to occur frequently enough. His former students and colleagues all had busy professional lives away from Lawrence that occupied most of their time. Retirement brought little opportunity to do what he had always enjoyed; now he had to rely mainly on his memories.

Later in the fall of 1947, Bunk received a diagnosis of diabetes caused by the hardening of his arteries. As the doctor tried to regulate Bunk's insulin levels, his clogged arteries caused great pain, making it difficult for him to move around. Homebound for the most part, now he could only occasionally entertain visitors.

In April of 1948 a visitor from the museum took a photograph of Bunk. Bunk's daughter Audrey said she never much cared for that photograph because she could see the pain in his face. For a man who had been sickly as a child, he actually possessed a strong constitution throughout his adult life. In spite of all the specimen preparation and taxidermy he had performed, the repeated exposure to the arsenic used to preserve the hides and the other toxic chemicals in the smelly old workshop, Bunk lived a long and physically active life. He quietly passed away at home on September 5, 1948, at the age of seventy-seven. According to the museum taxidermist at that time, George Young, who was present when he died, his breathing became slower and slower and finally ceased.

The next day Hall telegraphed Bunk's former students: "BUNKER PASSED AWAY SUNDAY FUNERAL TUESDAY THREE THIRTY NO FLOWERS." Two days after the memorial service, Hall wrote a letter to former students and friends of Bunk, giving them a more detailed account of his death. In the last paragraph of the letter, Hall captured just how much Bunk had meant to the museum family:

> Here at K.U., Bunk was the last one left of the early group that built up the Museum. We of younger generations still relied upon him for information concerning questions that arose about early-day specimens and records. In these matters and because the Museum just doesn't seem the same without him — and isn't the same — we miss him no end. However, if it's true that a man's good works live after him, Bunk will be with us in that sense for a long, long time to come.

Family and friends memorialized Bunk in early September. Unlike Dyche's large church funeral, which involved a lavish display of flowers, numerous speakers, and a long list of important dignitaries, Bunk's funeral was a simple affair, arranged by Clara, at a local mortuary. Those in attendance included family, a few friends from town, some former students from the Kansas City area, and staff and faculty from the museum. It was a straightforward and to-the-point service, just like the man whose life they celebrated. After the reading of standard bible passages and the singing of traditional funeral hymns, the small crowd dispersed. As the museum people and the family milled around after the service, they shared how Bunk's dedication to science, his appreciation of efficiency, and his natural gift as a problem solver made him an invaluable member of the museum. Everyone recognized how his generosity and humility inspired deep and lasting loyalty from those who knew and worked with him. Bunk's life gave testimony to the idea of how occasionally even the life of a seemingly ordinary person can have a profound effect on the lives of others.

A few days after the memorial service, the owner of the mortuary came by the Bunker home with Bunk's ashes. He drove Clara, Fedalma, and Audrey up to campus. The hot summers of Kansas don't usually break until mid-September, so in those days before air conditioning, classes for the fall semester at the University of Kansas had not started. The empty grounds at the university still looked like summer. The leaves on the trees showed signs of stress from hot breezes and little rain. The lawn of the campus, once a vivid green in early spring, was now dull and faded. When they climbed the stairs to the tower of Dyche Hall and reached the top, they were treated to that same magnificent view Bunk witnessed when he first discovered the top of Mount Oread in 1891. No other place for returning him to nature was more appropriate, given his affection for both the museum and the university. They stood in silence as the mortician offered the ashes to a southerly breeze that carried them gently toward the Kaw River.

Epilogue

Bunk lived during a time of great transition, from the post–Civil War period to two horrendous world wars; from western expansion and the Indian Wars to the advent of the automobile and the telephone. During his career at the University of Kansas much changed at KU and the natural history museum. For instance, Bunk's collection during his tenure grew significantly; both the ornithological and mammal collections quadrupled: from 27,000 to 110,000 and from 5,000 to 20,000, respectively. As the collections expanded, so did departmentalization and specialization. Today Bunk's collection of recent or modern vertebrates has evolved into separate collections; that is, ornithology, mammals, herpetology, among others, each with its own curator. With the increase in specialization came separation of responsibility and authority. Now curators would never think of collecting a specimen outside of their field like Bunk did when he collected the fossil of the gigantic sea serpent in 1911.

The years have increased museum security. In the early days at the museum, only a flimsy gate in a doorway discouraged unwanted visitors. Today electronic entrance cards and key codes limit access to the specimens. Additionally, the access door to the tower of Dyche Hall where Bunk's ashes were scattered was permanently sealed in the 1960s as a precaution against the occurrence of an event like the University of Texas tower shooting massacre.

As the burgeoning collection continued to grow, so did the need for safety. By the 1990s the sheer number of specimens stored in alcohol presented such a significant fire hazard that the City of Lawrence Fire Department informed the university that they would not fight any fire in Dyche Hall because of the danger to their firefighters. This admonition brought about the construction of a virtually fireproof addition to Dyche Hall to store the volatile wet specimens.

Much like when the museum of natural history outgrew the name "cabinet of natural history," today it goes by the name of the University of Kansas

Biodiversity Institute. The new name reflects a larger mission that emphasizes research and the interconnectedness of the Earth and all creatures and plants. Since Bunk's time the change in scientific technology has been phenomenal, with the advent of the internet, understanding of genomes and DNA, and satellites orbiting the world to assist with navigation and communications, television, and wireless communications. Today's scientists, in addition to collecting bones and skins, regularly collect, freeze, and analyze tissue samples to test for DNA and continue to identify new species on our planet. The old name of the natural history museum still survives, but only as the division of the Biodiversity Institute that deals primarily with the public displays, including the wonderful old *Panorama of North American Mammals.*

As the alliteration of "Bunk's boys" easily rolls off one's tongue, it is a reminder that during the early days in the museum of natural history, most graduate students were male. With the exception of the woman who wrote the poem in the 1935 alumni magazine article about Bunk and Sprague's reference to "two women" who had signed the bison bone, the natural sciences were pretty much the domain of men. Today that is not the case; the graduate students of the KU Biodiversity Institute are equally divided between the sexes.

In recent years the mounted animals in the panorama that Bunk helped construct have begun to show severe signs of wear and decay. Visitors can easily see cracks in many of the animal hides—predictable damage, considering they date back over 125 years to the 1893 Columbian Exposition. Prior to the 1930 renovation, the lack of climate control and years of ultraviolet rays from the skylight that initially existed over the panorama caused irreparable damage to the animals in the panorama. Recently the museum engaged a professional conservator to assess the entire panorama, including all the animals, plants, and the physical structure, and to clean the mounts and inventory all of the damage. The conservator noted that the *Panorama of North American Mammals* is an extraordinary treasure. Not only is it the largest and most comprehensive diorama in the world, it is one of the oldest, with enormous historical significance. The museum began to review its options, recognizing the importance of both preserving the valuable legacy and history of the panorama while making sure it reflects the scientific thinking and knowledge of the twenty-first century.

In 2018 Dyche Hall received a new tile roof and a remodeled top floor. That renovation removed the old walk-in vault that once housed Bunk's papers and field notes, although the steel door was left in place as a historical reminder. Not original to the old building, the vault was added with the 1930s renovation. To make it fireproof, its walls were constructed of hollow red clay tiles. During the demolition of the vault, the construction crew discovered a small glass test tube inside one of the tiles in the wall. A cork stopper sealed with wax perfectly preserved the contents: some cotton batting, like that used to pack bird specimens, a specimen tag containing a single word, "note," and a folded and rolled-up sheet of museum stationery. Carefully conserved, like most specimens in the museum, the handwritten note offered the following message: "To Whom It May Concern: The following persons were working on the scientific collections of this museum on this 18th day of May, nineteen hundred and forty." Written in Bunk's own hand were the following names: C. D. Bunker, Malcolm Brumwell, George Rinker, James Haupti, Klaus Abegg, and Paul Tiemeier. That small time capsule epitomized Bunk's life. In a quiet and unassuming manner, he gave recognition to those around him, meticulously making sure that the note would be preserved, and left it in a very private and obscure place outside of the limelight.

Historically, natural history museums take a unique and distinctive perspective toward the objects in the collection. In 1986 Professor John E. Simmons wrote in the foreword of a publication of the natural history museum at the University of Kansas, "Natural history museums tend to have extensive collections compared to other types of museums because their specimens must represent the entire range of variation within a species. In this regard, they are the opposite of art and history museums. These museums collect certain objects because they are unique — natural history museums collect them because they are not." In spite of all Bunk's accomplishments and the devotion he received from his students, in many ways Bunk was ordinary and common, like the objects of his beloved collection. But just because one's life is not unique or the subject of widespread celebrity does not make it less important or significant. In the final analysis, maybe the value of Bunk's story is that it resonates on a personal level with the average person more than all the stories of the important people in the history books.

Until a few years ago curators at the KU Museum of Natural History knew only that Bunk's cabin was located "7½ miles southeast of Lawrence." Given the use of GPS coordinates, that lack of precision seemed unusual in today's catalogs of birds and mammals. If anyone at the museum had tried using the county real estate records to find its location, searching for the name Bunker would have come up empty — the land where the cabin was located was owned by Clara's family. When its location was recently determined, the museum's catalog was updated. It turns out that today Bunk's cabin no longer stands. Only a few rocks from the foundation and a piece of rusted rain gutter survive. I know there will never be a marker memorializing Bunk's cabin and those woods will continue to consume any remnants of the structure, but at least I visited the exact spot where the cabin sat in the woods near Washington Creek. It was important to me so I could connect to the place where Bunk influenced so many of this country's best naturalists of the mid-twentieth century.

Bibliography

ARCHIVAL RESOURCES

KENNETH SPENCER RESEARCH LIBRARY, UNIVERSITY OF KANSAS LIBRARIES

University Archives

Brown, Barnum. Letter to H. H. Lane, October 6, 1931. Henry Higgins Lane Collection, PP 197, box 1, folder 2.

Burt, William. Letter to Deane Malott, June 27, 1942. Malott Papers, University of Kansas Chancellor's Office Records, 2/10.

———. Letter to Deane Malott, June 26, 1944. Malott Papers, University of Kansas Chancellor's Office Records, 2/10.

———. "News Letter No. 2," July 27, 1942. Copied to Deane Malott. E. Raymond Hall Papers, 513/3/6.

Goodwin, H. W. Letter to Deane Malott, March 10, 1942. Malott Papers, University of Kansas Chancellor's Office Records, 2/10.

Hall, E. R. Untitled internal memorandum, September 6, 1948. E. Raymond Hall papers, PP 167.

Hesse, Curtis. Letter to H. H. Lane, January 27, 1931. Henry Higgins Lane Collection, PP 197, box 1, folder 6.

Housholder, Vic. Letter to Deane Malott, March 12, 1942. Malott Papers, University of Kansas Chancellor's Office Records, 2/10.

K.U. News Bureau. Untitled press release, July 31, 1935. University of Kansas Natural History Museum records, RG 33/0/2.

Lane, H. H. Letter to Barnum Brown, October 9, 1931. Henry Higgins Lane Collection, PP 197, box 1, folder 2.

———. Letter to E. C. Case, January 31, 1931. Henry Higgins Lane Collection, PP 197, box 1, folder 3.

Malott, Deane. Letter to Charles D. Bunker, August 10, 1942. Malott Papers, University of Kansas Chancellor's Office Records, 2/10.

———. Letter to William Burt, June 29, 1942. Malott Papers, University of Kansas Chancellor's Office Records, 2/10.

———. Letter to H. W. Goodwin, March 12, 1942. Malott Papers, University of Kansas Chancellor's Office Records, 2/10.

———. Letter to Vic Housholder, March 16, 1942. Malott Papers, University of Kansas Chancellor's Office Records, 2/10.

Martin, Handel T. Letter to Charles D. Bunker and Theodore Rocklund, September 11, 1912. University of Kansas Natural History Museum records, RG 33.

Moore, R. C., and Bennet Allen. Agreement dated January 14, 1922. Ernest Lindley Papers, University of Kansas Chancellor's Office Records, 2/9.

Palmer, T. S. Notice to Charles D. Bunker, October 10, 1923. University of Kansas Museum of Natural History records, RG 33/1/12.

Snow, Francis Huntington. Letter to Jane Aiken, February 2, 1868. F. H. Snow Papers, University of Kansas Chancellor's Office Records, 2/6.

Kansas Collection

Burns, Thomas. Handwritten notes, "Charles (Pug) Saunders (Comanche)." Thomas L. "Tommy" Burns Collection, RH MS 1097.8.

Lawton, Keith. "Charles Saunders Decoys on the Kansas River." Thomas L. "Tommy" Burns Collection, RH 1907.6.

UNIVERSITY OF KANSAS NATURAL HISTORY MUSEUM

Bunker, Charles D. Handwritten field notes on Oklahoma Territory, 1901.

———. Handwritten field notes on Western Kansas, 2 vols., 1911.

———. Handwritten field notes on Alaska Territory, 1919.

DeLano, Raymond J. "The Diary of an Alaskan Trip," 1919.

PUBLISHED AND ONLINE SOURCES

Adams, Sam. Letter to C. D. Bunker, December 18, 1910. Dyche Papers, University of Kansas Archives. Quoted in Sharp and Sullivan, *Dashing Kansan*, 137.

Akeley, Carl Ethan. *In Brightest Africa*. Memorial ed. Garden City, NY: Doubleday, Page, 1920.

Andreas, A. T. *History of the State of Kansas*. Chicago: A. T. Andreas, 1883.

Anthony, D. R. "The Daily Times—The Stuffed Animal State." *Leavenworth Times*, December 17, 1982, 2.

Asma, Stephen T. *Stuffed Animals and Pickled Heads: The Culture and Evolution of Natural History Museums*. New York: Oxford University Press, 2001.

Bechtel, Stefan. *Mr. Hornaday's War: How a Peculiar Zookeeper Waged a Lonely Crusade for Wildlife That Changed the World*. Boston: Beacon, 2012.

Brownlee, Richard S. *Gray Ghosts of the Confederacy: Guerilla Warfare in the West, 1861–1865*. Baton Rouge: Louisiana State University Press, 1958.

Bunker, C. D. "The Birds of Kansas." *Kansas University Science Bulletin* 7, no. 5 (June 1913): 137–158.

———. "Habits of the Black-Capt Vireo (*Vireo atricapillus*)." *Condor* 12, no. 2 (March–April 1910): 70–73.

Cochran, Diane. "Historic Horse on Display in Kansas." *Billings (MT) Gazette*, June 23, 2005.

Colbert, Edwin H. *The Great Dinosaur Hunters and Their Discoveries*. New York: Dover Publications, 1984.

Commissioner of Indian Affairs. *Sixty-First Annual Report to the Secretary of the Interior*. Washington, DC: Government Printing Office, 1892. https://babel.hathitrust.org/cgi/pt?id=mdp.39015034627862.

Dary, David. *Lawrence, Douglas County, Kansas: An Informal History*. Lawrence, KS: Allen, 1982.

Dingus, Lowell, and Mark A. Norell. *Barnum Brown: The Man Who Discovered* Tyrannosaurus rex. Berkeley: University of California Press, 2010.

Duellman, William E. *Herpetology at Kansas: A Centennial History.* Ithaca, NY: Society for the Study of Amphibians and Reptiles, 2015.

Dwigans, Cathy M. *A Guide to the Museum of Natural History, The University of Kansas.* Lawrence: Museum of Natural History, University of Kansas, 1984.

"Dyche Museum to be Vastly Improved." *Graduate Magazine of the University of Kansas* 38, no. 1 (September 1939): 15.

Everhart, Michael J. *Oceans of Kansas: A Natural History of the Western Interior Sea.* Bloomington: Indiana University Press, 2005.

"Faculty." *Graduate Magazine of the University of Kansas* 41, no. 4 (1942): 12.

"Familiar Names, Faculty." *Graduate Magazine of the University of Kansas* 30, no. 1 (October 1931): 28.

Fargo, William G. "Auxiliary Gun Barrels for Collecting Bird Specimens." *Wilson Bulletin* 39, no. 4 (October–December 1927): 219–222.

"Finding Aid to the Ward C. Russell papers." Museum of Vertebrate Zoology, University of California, Berkeley, MVZA.MSS.0221. Accessed February 13, 2018. http://www.oac.cdlib.org/findaid/ark:/13030/c84b332g/.

Findley, James S., and J. Knox Jones, Jr. "Obituary: Eugene Raymond Hall: 1902–1986." *Journal of Mammalogy* 70, no. 2 (May 1989): 455–458.

Gregory, Joseph T. "Ruben Arthur Stirton (1901–1966), Fourth Director of UCMP." Museum of Paleontology, University of California, Berkeley. Accessed February 23, 2018. http://www.ucmp.berkeley.edu/about/history/rastirton.php.

Griffin, Clifford S. *The University of Kansas: A History.* Lawrence: University Press of Kansas, 1974.

Hall, E. Raymond. "In Memoriam: Charles Dean Bunker 1870–1948." Special publication. University of Kansas Museum of Natural History, 1951.

Hall, E. Raymond, and Ward C. Russell. "Dermestid Beetles as an Aid in Cleaning Bones." *Journal of Mammalogy* 14, no. 4 (November 13, 1933): 372–374.

Hartsock, D. Lane. "The Impact of the Railroads on Coal Mining in Osage County, 1869–1910." *Kansas Historical Quarterly* 37, no. 4 (Winter 1971): 429–440.

Hopewell Culture National Historical Park. "The Spanish Influenza Pandemic of 1918 at Camp Sherman." National Park Service. Accessed December 20, 2017. https://www.nps.gov/articles/influenza-at-camp-sherman.htm.

Hornaday, William T. *Taxidermy and Zoological Collecting: A Complete Handbook for the Amateur Taxidermist, Collector, Osteologist, Museum-Builder, Sportsman, and Traveller.* With W. J. Holland. 4th ed. New York: Charles Scribner's Sons, 1894.

Howard, Hildegard. "Notes and News." *Condor* 36 (September 15, 1934): 222–223.

"H. T. Martin, Curator of Paleontology, Passes." *Graduate Magazine of the University of Kansas* 29, no. 4 (January 1931): 12.

Hyder, Clyde Kenneth. *Snow of Kansas: The Life of Francis Huntington Snow, with Extracts from His Journals and Letters.* Lawrence: University Press of Kansas, 1953.

Jaffe, Mark. *The Gilded Dinosaur: The Fossil War Between E. D. Cope and O. C. Marsh and the Rise of American Science.* New York: Crown, 2000.

Kansas Sampler Foundation. "Council Grove, Santa Fe Trail National Historic Landmark." *8 Wonders of Kansas*. Accessed March 31, 2015. https://www.kansassampler.org/8wonders/historyresults.php?id=274.

Kellogg, Remington. "On the Retention of *Neotoma campestris* Allen as a Separate Subspecies from *Neotoma floridana baileyi* Merriam." *University of Kansas Museum Natural History, Zoological Series* 1, no. 1 (January 30, 1914): 3–6.

Kohl, Michael F., Larry D. Martin, and Paul Brinkman, eds. *A Triceratops Hunt in Pioneer Wyoming: The Journals of Barnum Brown and J. P. Sams: The University of Kansas Expedition of 1895*. Glendo, WY: High Plains, 2004.

Lanham, Url. *The Bone Hunters: The Heroic Age of Paleontology in the American West*. New York: Dover Publications, 1973.

Lazzarino, Chris. "Monsters of the Mind." *Kansas Alumni* 6, 2017, 25–29.

McCarthy, Erin. "The Flesh-Eating Beetles that Work at Natural History Museums." Mental Floss. September 11, 2015. Accessed February 28, 2017. http://mentalfloss.com/article/68184/beetle-work-natural-history-museums.

McQueen, Keven. "Carrie Nation: Militant Prohibitionist." *Offbeat Kentuckians: Legends to Lunatics*. Kuttawa, KY: McClanahan, 2001. Quoted in Sandra Bertrand, "American Spirits: A Look Back at the Prohibition Era," *Highbrow Magazine*, December 18, 2012. Accessed April 2, 2015. http://www.highbrowmagazine.com/1858-american-spirits-look-back-prohibition-era.

Mendota Centennial Committee. *Magnificent Whistle Stop: The 100-Year Story of Mendota, Ill.* Mendota: Mendota Centennial Committee, 1953.

Milgrom, Melissa. *Still Life: Adventures in Taxidermy*. New York: Houghton Mifflin Harcourt, 2010.

"The Mosasaurs." *Prehistoric Wildlife*. Accessed December 12, 2017. http://www.prehistoric-wildlife.com/articles/mosasaurs.html.

Museum of Vertebrae Zoology, University of California, Berkeley. "MVZ History." *Doing Natural History: The Journey of an Animal in the Museum of Vertebrate Zoology*. Accessed February 23, 2018. http://mvz.berkeley.edu/DoingNaturalHistory/history.html.

"Museum Reopened: Dr. Wetmore Pays Tribute to Snow and Dyche in Dedicatory Talk." *Lawrence Journal World*, June 7, 1941.

Niblack, Leslie G. "Geologist at Norman: To Make Another Extensive Territorial Trip." *Guthrie Daily Leader*, April 20, 1901.

Obituary of Lawrence V. Compton. *Washington Post*, February 2, 2003. Accessed July 22, 2018. https://www.washingtonpost.com/archive/local/2000/02/13/obituaries/23e53a6e-1f32-4bf4-88ee-b20b53f4ab7c.

Ortega y Gasset, José. *Meditations on Hunting*. Translated by Howard B. Wescott. Belgrade, MT: Wilderness Adventures, 2007.

Osborn, Henry Fairfield. *Biographical Memoir of Edward Drinker Cope, 1840–1897*. Biographical Memoirs of the National Academy of Sciences, vol. 8. Washington, DC: National Academy of Sciences.

Palmer, F. D. "The Kansas Exhibit of Mounted Specimens of the State." *Scientific American* 69, no. 3 (July 15, 1893): 41.

Penner, Catherine. "Mr. Charles D. Bunker, Trainer of Scientists." *Graduate Magazine of the University of Kansas* 34, no. 4 (January 1936): 4–6.

"Plesiosaurus." *Prehistoric Wildlife*. Accessed December 12, 2017. http://www.prehistoric-wildlife.com/species/p/plesiosaurus.html.

Rempel, Janae. "A Geological History of Meade County According to Claude Hibbard." *Old Meade County: Sharing the History of Meade County, Kansas On-line*. Meade County Historical Society. Accessed February 22, 2018. http://www.oldmeadecounty.com/hibbard.htm.

Rogers, Katherine. *The Sternberg Fossil Hunters: A Dinosaur Dynasty*. Missoula, MT: Mountain, 1999.

"Scranton History." City of Scranton, Kansas. Accessed March 31, 2015. http://www.scrantonks.com/history.html.

Serafini, Anthony. *The Epic History of Biology*. Cambridge, MA: Perseus, 1993.

Shankel, Carol, and Barbara Watkins. *Dyche Hall: University of Kansas Natural History Museum, 1903–2003*. Lawrence: Historic Mount Oread Fund, Kansas University Endowment Association, 2003.

Sharp, William, and Peggy Sullivan. *The Dashing Kansan: Lewis Lindsay Dyche: The Amazing Adventures of a Nineteenth-Century Naturalist and Explorer*. Kansas City, MO: Harrow Books, 1990.

Sheiff, Judy. "A Brief History of the Kansas Biological Survey." *BIOTA the University of Kansas Biological Survey* 3 (Winter 1995): 1.

Shor, Elizabeth Noble. *Fossils and Flies: The Life of a Compleat Scientist, Samuel Wendell Williston (1851–1918)*. Norman, OK: University of Oklahoma Press, 1971.

Sprague, James M. "James M. Sprague." In *The History of Neuroscience in Autobiography*, edited by Larry R. Squire, vol. 1, 498–526. Washington, DC: Society for Neuroscience, 1996. Accessed July 17, 2018. https://www.sfn.org/about/history-of-neuroscience/autobiographical-chapters.

Sternberg, Charles Hazelius. *The Life of a Fossil Hunter*. New York: H. Holt, 1909.

"Storm Visited Lawrence." *Kansas City Times*, April 13, 1911.

Taft, Robert. *The Years on Mount Oread: A Revision and Extension of Across the Years on Mount Oread*. Lawrence: University Press of Kansas, 1955.

Thomas, Robert McG., Jr. "Deane Malott, 98, Educator Who Ran Two Universities." *New York Times*, September 13, 1996.

"University Notes." *Lawrence Daily World*, February 10, 1903.

Whitmore, Frank C., Jr. "Remington Kellogg 1892–1969: A Biographical Memoir." Washington, DC: National Academy of Sciences, 1975. http://www.nasonline.org/publications/biographical-memoirs/memoir-pdfs/kellogg-remington.pdf.

Wilson, John Andrew. "Memorial to Claude William Hibbard, 1905–1973." *Geological Society of America, Memorials*, vol. 5 (1977): 1–8. Accessed August 10, 2018. http://www.geosociety.org/documents/gsa/memorials/v05/Hibbard-CW.pdf.

Index